GREEN TECHNOLOGY AND DESIGN FOR THE ENVIRONMENT

GREEN TECHNOLOGY AND DESIGN FOR THE ENVIRONMENT

Samir B. Billatos
Nadia A. Basaly
University of Connecticut
Storrs, CT

Taylor & Francis
Publishers since 1798

USA	Publishing Office:	Taylor & Francis
		1101 Vermont Ave., N.W., Suite 200
		Washington, DC 20005
		Tel: (202) 289-2174
		Fax: (202) 289-3665
	Distribution Center:	Taylor & Francis
		1900 Frost Road, Suite 101
		Bristol, PA 19007-1598
		Tel: (215) 785-5800
		Fax: (215) 785-5515
UK		Taylor & Francis, Ltd.
		1 Gunpowder Square
		London EC4A 3DE
		Tel: 071 583 0490
		Fax: 071 583 0581

GREEN TECHNOLOGY AND DESIGN FOR THE ENVIRONMENT

1 2 3 4 5 6 7 8 9 0 BRBR 9 8 7

This book was set in Times Roman. The editors were Christine Williams and Elizabeth Dugger. Cover design by Michelle Fleitz.

A CIP catalog record for this book is available from the British Library.
∞ The paper in this publication meets the requirements of the ANSI Standard Z39.48-1984 (Permanence of Paper)

Library of Congress Cataloging-in-Publication Data

Billatos, Samir B. (Samir Botros).
    Green technology and design for the environment / Samir B.
Billatos, Nadia A. Basaly.
        p. cm.
    Includes bibliographical references and index.

    1. Design, industrial—Environmental aspects. 2. Green
technology. I. Basaly, Nadia A. II. Title.
    TS171.4.B54 1997
    628—dc20   96-30958
                                                                                    CIP
    ISBN 1-56032-460-0 (case)

CONTENTS

Part II   Design Examples

7   DESIGN FOR ASSEMBLY AND DISASSEMBLY   131

8   DESIGN FOR ENVIRONMENTALLY FRIENDLY VEHICLES   173

9 DESIGN FOR MANUFACTURING PROCESS IMPROVEMENT

PREFACE

This book is intended to change your approach to the practical application of engineering. Recent developments in concurrent engineering have done this with great success by improving the processes or *methodologies* of design and manufacturing. Shorter development cycles, lower rejection rates, and improved quality are visible benefits of these new methods, which are now revolutionizing production practices through out the world. Green engineering provides a complementary focus by directing these new methods toward a specific purpose. It thus has only recently become practical to attempt the complex and ambitious goal of green engineering, which is: to define and internalize at the product design stage those "external" parameters that affect the overall environmental impact of a product: to eliminate toxic pollutants; to minimize waste and depletion of resources; and to accomplish this, in many cases, with a net cost savings.

Environmental regulations, to the extent that they affect the cost of products, are a relatively new development. Few people seriously discount this basic need, but many will debate the levels at which they are proscribed, citing the cost as an unfair burden in a highly competitive economy. Regulation is of course a necessary step, but unfortunately provides no real incentive for *proactive* improvement, for regulation is still interpreted as a license to avoid the external costs of environmental impact. In contrast to this position, recent experience has demonstrated that systematic elimination of waste and environmental impact can provide net economic and strategic benefits in higher quality products; more efficient operations; and the goodwill of an informed public that will expect a cleaner and healthier environment. As regulations become more demanding, and the public more aware and concerned, this incentive for green engineering will become increasingly apparent.

We hope to exploit this new approach to the maintenance of environmental quality. Companies now recognize that "end-of-pipe" regulatory compliance will ultimately become an expensive liability. Reprocessed materials and remanufactured products will become cost-saving strategies. Competitive survivability will demand systems-level solutions to the problems of environmental quality: toxics control, waste minimization, and resource management, through a fundamental redesign of products and processes. These requirements clearly pose new constraints to the already complex tasks of design and manufacturing. It is only with the latest tools of concurrent engineering and the ability to model all phases of cost and performance, prior to production, that one can hope to successfully contend with this level of complexity.

This book is intended to be an introductory reference to green technology and design for the environment (GTDFE) for working professionals as well as a basic text for graduate course work in engineering. With this objective in mind, the book is divided into two parts covering all facets of GTDFE. Part I consists of Chapters 1 to 6 and defines the problems the world faces today on the road to environmentally safe products. In Chapter 1, the basic subject of green engineering is introduced and defined. This subject has a broad interdisciplinary scope and is applicable to every phase of design and manufacturing. In Chapter 2 the basic problems and magnitude of waste and the need to revise current design practices are introduced. Chapter 3 is an attempt to summarize the monumental body of information on past examples of environmental damage, the growing body of regulations, and the newly emerging ethic of stewardship by some more progressive members of industry.

Chapters 4 to 6 present the methods of solution to these problems. They begin a systematic outline of criteria and methods for environmentally conscious engineering. These methods are presented in consistent format of "design for X" philosophy, where "X" implies a specific form of optimization, and extend to a thorough treatment of several important categories of optimization. Chapter 4 begins a general introduction and guidelines to designing for the environment, while Chapters 5 and 6 continue this theme with an in-depth treatment of recycling issues, especially plastics.

Part II consists of Chapters 7 to 12 and provides an encouraging number of examples and cases studies from industry. Chapter 7 provides an environmentally friendly approach to the subjects of assembly and disassembly, while Chapter 8 continues this theme with the presentation of electric vehicles as a successful endeavor of environmentally friendly products. Chapter 9 presents a general outline of manufacturing processes from an environmental perspective, and identifies a variety of new alternatives that waste less material and result in substantially less pollution. Knowledge of the adverse environmental characteristics of various processes in the design stage is an essential part of environmentally conscious engineering.

Chapters 10 to 12 conclude this book with some analyses and design tools for continuous improvement of quality products and processes. Chapter 10 is a practical application of designing for quality of environmentally safe cooling devices for cutting tools. The decision-making tools that are essential to accurate and creative design synthesis are outlined in Chapters 11 and 12. These methods will become the primary workhorses of environmentally conscious engineering, for they provide a much-needed degree of order to the expanded and complex scope of environmentally conscious design.

We hope that this strategy can provide the basis for a new ethos. Competitive production can exist within a clean environment. Through creative application of green engineering we can maintain a high level of environmental quality at an affordable price—and material comforts need not come at the expense of our environment.

We finally wish to thank and acknowledge the love and support of our family and our three children, Christine, Lydia, and Ehab. We also extend our sincere thanks to our students for their critical review of few chapters and for their valuable suggestions. In particular, we thank Larry Grigely for all his efforts in the first three chapters, along with William Tereshkovich and Steve Formella. We thank the publisher for an interest and understanding of the need for books in the green technology area.

PART
ONE

THEORY

ONE

GREEN ENGINEERING AND ENVIRONMENTALLY CONSCIOUS MANUFACTURING

Environmental safety is making its mark on the population of the world, specifically the world's industry. The engineering and manufacturing processes of a product play a major role in impacting the environment either positively or negatively, depending on the process. This chapter explains why such processes need to be environmentally friendly in order to succeed in the public eye.

1.1   INTRODUCTION: WHAT IS GREEN ENGINEERING?

Green engineering originated in the early 1970s when the environmental movement drew political attention to the high levels of consumption and waste needed to sustain the industrialized world.   During that time much attention was directed to the development of alternative energy sources and conservation measures due to a drastic increase in world oil prices.   This period also marked the beginning of a trend in the enactment of legislation and regulations to control the most damaging effects of industrial pollution.   Since then, the effects of humans and technology on our natural environment have become a topic of much research in many areas of science, engineering, and industry.   This body of work is now becoming recognized in its own right and when presented in a structured format, as the interdisciplinary subject of *green engineering.*[1]

Green engineering is a systems-level approach to product and process design where environmental attributes are treated as primary objectives or opportunities, rather than as simple constraints, and emphasizes the legitimacy of environmental objectives as consistent with the overall requirements of product quality and economy

---

1. "Green engineering," also known as "environmentally conscious manufacturing," or "green technology."

[1]. These environmental opportunities become most visible when evaluated within the context of the *system life cycle,* and can be easily overlooked when design tasks are confined within the standard domain of specific product requirements.

1.2 GOALS OF GREEN ENGINEERING

Green engineering involves four general goals, which can improve both environmental quality and the economics of production when implemented in the product design stage. These goals are as follows:

1.2.1 Waste Reduction

Waste reduction measures are generally justified on the basis of strict financial analysis without concern for the added environmental benefits. Economic benefits result from reduced budgets for material, waste handling equipment, and labor. Efficient material usage also results in other intangible benefits such as worker awareness and more accurate levels of material accounting. This is consistent with advanced operational strategies such as total quality management (TQM) and just-in-time manufacturing (JIT).

In more general terms, waste reduction also involves product design for such things as extended durability, which is often considered a measure of product quality, and the use of less net material, since in many cases lightweight products are considered more desirable. Environmental benefits can range from modest to sizable reductions in waste volume, landfill capacity, and transportation impact. These environmental benefits can provide secondary economic benefits that are often overlooked in preliminary financial analyses.

1.2.2 Materials Management

Materials management involves activities that lead to the recovery of materials or finished components for reuse in their highest value-added application. This involves making a material useful while minimizing the amount of added processing needed to effect recovery. Materials management also applies to the flow of hazardous materials with the intent of minimizing the associated exposure to product liability. These activities can be summarized by the following three categories.

Design for recycling (DFR). Applies to the cost-effective reuse of materials and whole components. Current design practices ignore the need to support the product disposal stage or the potential for reusing materials and components. New DFR methods will consider material and component life cycles [2] with the goal of maximizing the reusability of materials and components.

Design for disassembly (DFD). Applies to the use of assembly methods and configurations that allow for cost-effective separation and recovery of reusable components and materials. Significant savings will be achieved when disassembly activities are considered in the design stage. This will include better selection of materials, more specific identification of component materials, and assembly methods that provide for more efficient and possibly automated disassembly.

Toxics management. Applies to the elimination or control of toxic materials that are an intrinsic part of the product. Examples include such items as cadmium in batteries or lead solder in printed circuit boards. These materials pose an increasing threat to public health due to potential land and groundwater contamination after eventual disposal. Future regulations will likely require control of such materials. This topic is also considered within the context of pollution prevention but is mentioned here because of the direct association with the management of useful materials.

Materials management provides economic benefits by providing sources of resource-intensive materials and components at lower costs than could be had from virgin material or new component sources. The energy and resources "contained" within existing materials is expected to increase as products become increasingly sophisticated. The goal of effective material reprocessing currently presents a significant challenge to designers due to the range of materials and assembly methods used today. Toxics management provides an important economic benefit by eliminating expensive handling costs and, more importantly, the risk of environmental liability. This is becoming an important issue for an increasingly sophisticated and more environmentally conscious consumer public, and is a growing concern to financial institutions involved in the financing of product development and manufacture.

The environmental benefits of materials management result from a reduction in waste volume and landfill impact and, in the case of toxics management, a significant reduction in hazardous contamination of land and groundwater. The ultimate, although distant, goal of materials management is a closed-loop material cycle. The logic for this derives from the eco-development model of green engineering, and is presented in Chapter 3.

1.2.3   Pollution Prevention

The goal of pollution prevention is to eliminate the use of manufacturing processes that generate pollution. This can be achieved by either; re-design of processes to eliminate the production of harmful by-products, or; the redesign of products to eliminate the need for those processes that generate harmful by-products. These internalized methods differ from *pollution controls*, a term that refers to the treatment of harmful by-products after they have been created. This is also commonly known as the "end-of-pipe" (EOP) solution. Many production processes were developed prior to the enactment of environmental regulations and were adapted after the fact to comply with regulations and avoid the expense of regulatory fines. The operating cost of these methods is often poorly characterized prior to implementation. Pollution prevention generally requires capital investments but is generally less costly overall when compared to pollution control measures. Pollution control costs are expected to increase as future regulations become more progressive and continue to demand reduced levels of emissions.

Economic benefits of pollution prevention result from the reduced and more stable pollution compliance costs; the reduction or elimination of environmental damage liability; the reduced risk to worker health and safety; and the "good will" of customers and the public.

The environmental benefits are obvious: a healthier and less polluted environment.

1.2.4   Product Enhancement

In many cases, the function of a product can be significantly enhanced by the inclusion of features that result in less waste and pollution during use throughout its operating life. This includes systems that reduce the need for resources during operation, such as energy conserving appliances, washing machines that use less water, and intelligent traffic control systems.

Creative designers of the future will find many new profitable opportunities for application of environmental product enhancement. Economic benefits result from lower operating costs, and the strategic benefits of increased demand for more economical, higher quality products by an increasingly sophisticated and environmentally conscious public. Environmental benefits result from a reduction in waste and demand for resources over the life of the product.

Ideally, the application of green engineering principles will lead to an entirely new perspective on engineering in general, with the underlying goal of meeting the complex material needs of society in sustainable harmony with the life-giving forces of our natural world. This philosophy differs from current practice in many ways. Many present methods of design and manufacture are descended from a time when resources were plentiful, and the adverse environmental effects of new materials and high volume production were of incidental concern. Product performance and ease of production were the dominant criteria and established the requirements for design and manufacturing [3–5]. As the effects of environmentally harmful practices became evident, social pressures responded by placing restrictions, in the form of regulations, on the amount of environmental damage that could be legally permitted. These restrictions are based on what is now referred to as the *environmental protection* model, and represent the first stage in the evolution of environmentally progressive actions.[2]

Beneficial as it may be, the environmental protection model suffers from serious limitations due to its "external" nature. It focuses primarily on harmful by-products (not the products per se) and relies almost exclusively on the instrument of regulation. This has unfortunately lead to a common and erroneous mind-set: that *environmental quality results from compliance with regulations,* and that this is always in conflict with the profitable operation of business.

Green engineering is based on the more progressive environmentally

---

2. See Chapter 3 for the five paradigms of environmental regulation.

conscious models of *resource management* and *eco-development*; terms that are explained in greater detail in later chapters of this book. For now, the primary distinction is that systematic internalization of environmental requirements in the design stage can lead to less waste, more efficient methods of production, strategic opportunities for product development, and a healthier environment.

1.3   WHO NEEDS GREEN ENGINEERING?

Practically all production-oriented industries could benefit from the application of green engineering. This includes operations within the scope of any system life cycle, including raw material producers, manufacturers, product users, recyclers, and waste handlers. In general, the most intensive and wasteful industries are those at the early stages of the production cycle (75% of all manufacturing waste in 1990 was generated by; chemical, petroleum, paper, and primary metals industries [6]). This by no means limits the focus of green engineering to these producers, since the *manufacturers who use their finished materials, and can select the types and amounts, have a direct influence on the overall environmental impact that must be exchanged to ultimately produce a given product.* Thus, every kilogram of metal *not used* in production of a product represents a proportional savings of environmental impact: mine tailings, heavy metals used in ore processing, coal burning, other energy consumption, etc.

This effect has far reaching implications. In many cases, personnel who may be far removed from physical production processes are unaware of the waste-generating impact of their design choices. For example, a computer programmer may indirectly generate a significant amount of waste paper with a poorly designed report format used by a large number of people. Furthermore, paper manufacture is a pollution-intensive process, generating millions of tons of contaminated waste water annually. A simple change in the format of a report; at *essentially no cost,* could eliminate an extraordinary amount—literally tons of annually generated pollution!

In a similar real-world example, 2000 lb of copper cable was replaced by 65 lb of fiber optic cable. Copper production generates toxic heavy metal wastes such as arsenic, and requires destruction of large land areas by strip mining. The fiber production incurred only minimal impact, and consumed only 5% of the energy required for the copper.

These examples of green engineering are, in fact, simple, system-level solutions to the problems of waste and pollution. As opportunities for these methods of solutions become more common, the public will recognize and legitimately expect that environmental concerns be given high priority. In the interest of a sustainable future, we as an intelligent society need green engineering.

1.4   WHAT IS ENVIRONMENTALLY CONSCIOUS MANUFACTURING?

Environmentally conscious manufacturing (ECM) is the development and application of cleaner, more efficient manufacturing methods to achieve the goals of green engineering. This includes a systematic characterization of manufacturing processes for various measures of resource conversion efficiency, types and amounts of generated pollution, and the degree to which materials may be most effectively reused. This must include all known processes and secondary operations that are required to produce a given category of product. In many cases, secondary operations such as cleaning and painting are primary sources of pollution. Many new processes are currently being developed to replace pollution sources and to provide efficient reuse of materials. A summary of this information is presented in later chapters of this book.

The purpose of this characterization is to provide the means for evaluating the environmental quality of product manufacture and to conduct comparative trade studies of environmental impact at early stages in the design process. This sort of approach has shown to be successful for other manufacturing characterization methodologies, and is sometimes called the design for "X" method, where X implies a specific form of optimization, such as zero defects or automated assembly.

The term "design for the environment" (DFE) has been used to identify design practices where the environmental effects of product function and manufacture are optimized in the design stage. This new strategy is the primary focus of green engineering. When coupled with other DFX strategies, DFE can improve product quality through more effective methods of materials management and the elimination of wasteful and polluting manufacturing processes. Figure 1.1 shows and Table 1.1 lists a number of common processes that have been revised or eliminated through the use of ECM.

10.9 Billion Tons Total Waste (RCRA)

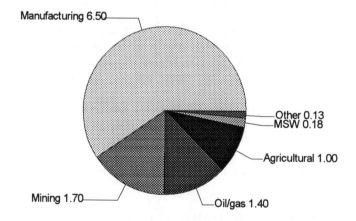

Figure 1.1 Manufacturing is the largest contributor to waste generation.

Table 1.1 Types of Waste Generated by Manufacturing and Alternatives

Types of Waste *Generated by Manufacturing*	*Alternatives*
1. Volatile organic cleaning fluids.	1. Use closed loop cleaning processes. Use environmentally solvents.
2. Process waste water.	2. Redesign manufacturing processes for closed loop washing and cooling.
3. Machining lubricants.	3. Use coolant-less machining processes.
4. Material removal scrap.	4. Use "Scrapless" forming processes.
5. Low yield manufacturing processes.	5. Improve process yield through quality control, design of experiments, and SPC.
6. Toxic materials in municipal waste stream.	6. Contain toxics with "Take Back" regulations. Eliminate or minimize toxics in design stage.
7. Energy consumption.	7. Design efficient, "Energy Smart" appliances. Use lightweight materials for transportation.
8. Material disposal at end of product life.	8. Design products for disassembly and re-use of materials and components.

1.5   WHO NEEDS ENVIRONMENTALLY CONSCIOUS MANUFACTURING?

Any business involved in the manufacture of materials or products throughout the system life cycle can benefit from the implementation of ECM. Table 1.2 lists several production processes and the possible benefits to be gained. As previously mentioned, the early stages of the production life cycle tend to generate the highest proportion of wastes. Improvements in these processes are in many cases industry specific and generally require large capital expenditures for new plant equipment. Examples of this include chemical production and primary metals refining. These industries will feel increasing pressure to modernize their operations as green engineering becomes more widely recognized, since DFE applications based on life-cycle analyses will strongly favor the use of materials produced using ECM methods.

Product fabrication industries also generate a significant net volume of waste and pollution, much of which can be drastically reduced or completely eliminated. Materials reprocessing industries are also included in this category. These areas will realize the most immediate benefits through the application of ECM, and will therefore receive primary attention through much of this book.

Table 1.2. Green Engineering and Stages of the Product Life Cycle

	MATERIAL EXTRACTION Example: Mining	MATERIAL PROCESSING Example: Smelting	MANUFACTURING Example: Auto Parts	USE Example: Transportation	WASTE MANAGEMENT Example: Recycling
TYPES OF WASTE	• Mine Tailings • Transportation Energy	• Waste Water • Toxic Metals • Process Energy	• Material Scrap • Cleaning Fluids • Coolants • Volatile Paints	• Air Pollution • Energy Consumption • Toxic Fluids	• Disassembly Time • Unrecoverable Materials • Unrecoverable Components
SOLUTIONS BY GREEN ENGINEERING	• Reduced Consumption	• Reduced Consumption • Toxics Containment	• Eliminate Coolants • Closed Loop Cleaning Processes • "Scrapless" Processes	• Increase Fuel Efficiency • Recycle Fluids • Increase Durability • Increase Maintainability	• Design For Disassembly • Design For Recycling • Design For Remanufacture • Standardize Material Identification

1.6 SUMMARY

Green engineering is a newly emerging concept for managing the environmental effects of product manufacture by including these important parameters at the early design stage. Waste reduction, materials management, pollution prevention, and product enhancement are the four primary goals for its implementation.

Environmentally conscious manufacturing is the application of green engineering at the industrial and production stage with specific emphasis on extending existing methodologies such as design for "X" (DFX) product improvement, improved methods for process control, such as statistical process control (SPC), and improved material tracking and control.

Green engineering has great potential for changing currently perceived conflicts between profitability and environmental quality into a "win–win" opportunity. As companies begin to compete on this new ground, cleaner and more efficient production processes will become more profitable and strategically necessary for survival.

1.7 BIBLIOGRAPHY

1.   *Green Products By Design: Choices for a Cleaner Environment*, OTA-E541, U.S. Congress, Office of Technology Assessment, Washington, DC, 1992.
2.   *Life Cycle Assessment: Inventory Guidelines and Principles*, EPA/600/R-92/245, Risk Reduction Research Laboratory Office of Research and Development, U.S. Environmental Protection Agency, Cincinnati, OH, February 1993.
3.   Reijnders, L. *Environmentally Improved Production Processes and Products: An Introduction.* Kluwer Academic Publishing, New York, 1996.
4.   Shina, S.G. *Concurrent Engineering and Design for Manufacture of Electronics Products*, Van Nostrand Reinhold, New York, 1991.
5.   Altung, L. *Manufacturing Engineering Processes*, 2nd edition, Marcel Decker, New York, 1994.
6.   *Managing Industrial Solid Wastes from Manufacturing, Mining, Oil, and Gas Production, and Utility Coal Combustion*, OTA-O-BP-82, U.S. Congress, Office of Technology Assessment, Washington, DC, 1992.

1.8 PROBLEMS / CASE STUDIES

1.1   Green engineering is a stumbling block to many industries. The reason is their misunderstanding of the potential benefits if properly implemented. In your opinion, would you classify green engineering as a quick fix program or a long-term process? Is it capital intensive or communication intensive? How can you use your answer to convince these industries?

1.2   In a green engineering environment, are design for "X" methodologies (DFX as defined in Section 1.4) up-front activities or downstream activities. Would your answer lead to standardization of our design and manufacturing practices? For example, would your answer affect the number of parts of a product, the manufacturing time and cost of the product, the serviceability and testability of the product, etc.?

1.3 Industry that opposes the implementation of green engineering claims that its delay in practicing green engineering requires a substantial organizational culture change. Some others claim that green engineering is a design process change. Can you help these industries to distinguish the impact of green engineering on their culture or design practice.

1.4 Examine Table 1.2 and provide additional industries that generate waste. For each industry, provide the waste origin, waste type, and the pollution prevention and recycling methods. Examine such industries as (a) fabricated metal industry such as metal-working fluids in machining, metal wastes dust sludge in machining, and solvents in parts cleaning, (b) fiberglass-reinforced composite plastics industry such as rejected and excess raw material in raw material purchasing and unloading, gelcoat resin and solvent overspray in fabrication, and air emissions in curing, (c) steel industry such as scrap steel in receiving, slag in foundry work, and oils greases in hot and cold rolling processes, etc.

1.5 Is there a conflict in an environmentally conscious manufacturing environment? If there is, is it healthy and desirable? Does environmentally conscious manufacturing require management commitment and participation? In what capacity? What is the management role in this case?

TWO

INFRASTRUCTURE, REGULATIONS, AND THE WASTE STREAM

One man's junk is another man's treasure.—Anonymous

In this chapter, the current terminology used to characterize the waste stream, with special emphasis on waste and by-products of manufacturing operations, is introduced. The purpose is to identify the *resource utility* of given waste products as potential opportunities. The chapter closes with a discussion of existing methods of waste handling, their limitations, and the opportunities for environmentally conscious engineering within a newly emerging industrial infrastructure.

2.1   INTRODUCTION

The concept of "waste" is known to us in a variety of ways. A useful meaning for our purpose implies *less than ideal utilization of any resource*, such as time, energy, or materials. From this perspective the characterization of wastefulness requires, in the most basic sense, some *model of ideal utilization*. Waste, wherever it can be identified, indicates less than ideal utilization. Equally important is the degree of liability associated with a given waste. Hazardous wastes, by their very nature, have *negative resource utility* because they incur significant expense from accident liability and the costs associated with insurance, disposal, and storage [1,2].

Accurate characterization of this nation's waste stream is an enormous undertaking that has only recently been attempted by government agencies. These agencies assist in legislative and regulatory policymaking. Few private organizations derive benefit in collecting this type of data. Much of the information gathered is presented within very broad and often ambiguous formats proscribed by regulatory agencies. Of this information, Table 2.1 provides a summary from which three important conclusions can be made. First, the amount of waste material generated in the United States is staggering. In 1991 industries generated, on average, over 40 tons of waste matter per capita. (This figure only includes forms of waste that the U.S. Environmental Protection Agency [EPA] monitors and it does not account for significant additional sources for which there are no reliable estimates [3-7].) Second, clearly wasteful operating practices are relatively common. Finally, either these practices are accepted as a cost of business, or they occur because they are invisible. For a variety of reasons, the extent and cost of waste are not fully recognized.

Table 2.1. Estimated Amounts of Solid Wastes by Manufacturing Industry

Industry	Amt 10^6ton	EPA toxicity assessment and concerns
Pulp and paper	2,250	Moderate. Organics, heavy metals, sulfates & some dioxins present.
Primary iron and steel	1,300	High. Low discharge pH with significant amounts of heavy metals.
Inorganic chemicals	920	High. Low control of on site hazardous waste disposal.
Stone, clay, glass, & concrete	622	Low. Most wastes are inert.
Food & food products	374	Low. Most wastes are bio-degradable.
Textiles	254	Low. Low organics and heavy metals, but little analytical data exists.
Plastics and resins	181	High. High in organic solvents & toxic
Petroleum refining	169	High. High in sulfides, ammonia, phenols & other toxic organics.
Fertilizer & agricul chemicals	166	High. High in toxic organics and heavy metals.
Primary nonferrous metals	67	High. High levels of heavy metals.
Organic chemicals	59	High. High levels of toxic organics.
Water treatment	59	Low. Some traces of heavy metals.
Rubber & miscellaneous products	24	High. Includes plastic resins and pigment compounds.
Transportation equipment	13	High. High levels of oils, heavy metals & toxic organics
Leather & leather products	3	Moderate. Significant Chromium, although in benign (trivalent) form.
Miscellaneous	64	
Total	6,524	

Businesses may often overlook wasteful operations because they fall within the limits of common practice. In this sense there is a *homeostasis to the industrial metabolism*; a term normally used to describe the processes of living organisms. Companies can function successfully if their resource utilization falls within a range of average nominal efficiency.

This average nominal efficiency is typically gauged by their competitors' performances. Aside from this, the impetus for competitive advantage can then focus in a variety of other areas. Profitability can be maintained by adjusting prices or by devising other forms of market strategy to stimulate consumption. This practice sets the standard for accepted levels of efficiency. Generally profit motives alone are not sufficient mechanisms for seeking high degrees of operating efficiency.

When the average nominal efficiency of resource utilization changes, for example, due to the development and adoption of newer technology by an entire industrial segment, operational efficiency becomes an issue of strategic importance, or perhaps, a matter of survival. New technologies are developing rapidly. Much of this development has been stimulated by a relatively new and sometimes demanding requirement: the need to comply with environmental regulations.

2.2   THE WASTE STREAM

Waste is generated in all stages of production, product use, and disposal [8,9]. Since industries control the largest share of this cycle, it is no surprise that they are the major waste generators: over 98% of all waste is generated by industry. The fractional remainder is generated by the public as *municipal solid waste* (MSW). Most of the terms for defining waste derive from regulations that control disposal rather than the actual waste composition per se. Waste will be described in terms of its modes of disposal, and to a lesser extent by the regulations that control disposal. These modes are air, water, and land.

2.2.1   Air Pollution

Gaseous wastes are a significant by-product in many manufacturing processes [10]. Primary sources are *criteria pollutants, volatile organic compounds* (VOCs), and *particulates*. Criteria pollutants are products of combustion, VOCs are vapors released during painting and cleaning processes, and particulates are dust from casting and metal finishing processes. These and other air pollution sources are listed in Tables 2.2 and 2.3. In almost all cases these wastes are classified as *pollutants* and are strictly controlled by regulations, which will progressively become more restrictive in the future. A number of commonly used volatile organic compounds are listed in Table 2.4.

Table 2.2. Air Pollution Sources in Manufacturing

Pollution Source	Pollutant Type
Internal Combustion Engines Fossil Fueled Power Generation Waste Incineration	Criteria Pollutants: Primary Products of Combustion
Coating Processes Paint Mixing and Application Fabricated Metals Fiberglass & Reinforced Composites Parts Cleaning Metal Finishing CFC Refrigerant Systems Printed Circuit Board Processing	Volatile Organic Compounds: Used in Cleaning & Coating Processes
Casting Processes Bulk Materials Handling Textile Handling Grinding and Abrasive Processing	Dust and Particulate with Possible Toxic Content

Table 2.3. Criteria Air Pollutants

Pollutant	Source	Effects
Sulfur Dioxide	Internal Combustion Engines and Coal Fired Power Plants	Adverse To Respiratory Health, Source Of Acid Rain
Nitrogen Oxides	Internal Combustion Engines	Adverse To Respiratory Health
Carbon Monoxide	Internal Combustion Engines and Coal Fired Power Plants	Adverse To Respiratory Health
Ozone	Photo-Chemical Reactions Of Other Pollutants	Adverse To Respiratory Health
Particulate	Internal Combustion Engines and Coal Fired Power Plants	Adverse Respiratory Health, Possible Carcinogen
Lead	Gasoline Engines Using Leaded Gasoline	Adverse To Health, Possible Nerve And Liver Damage

Table 2.4. Some EPA-Listed Hazardous Waste Solvents

Solvent	EPA Hazardous Waste Number
Benzene	F005
Carbon Disulfide	F005
Carbon Tetrachloride	F001
Chlorobenzene	F002
Cresols	F004
Cresylic Acid	F004
O-Dichlorobenzene	F002
Ethanol	D001
2-Ethoxyethanool	F005
Ethylene Dichloride	D001
Isobutanol	F005
Isopropanol	D001
Kerosene	D001
Methyl Ethyl Keytone	F005
Methylene Chloride	F001, F002
Naphtha	D001
Nitrobenzene	F004
2-Nitropropane	F005
Petroleum Solvents	D001
Pyridine	F005
1,1,1 Trichloroethane	F001, F002
1,1,2 Trichloroethane	F002
1,1,2 Tetrachlorethylene	F001
Toluene	F005
Trichloroethylene	F001, F002
Trichlorofluromethane	F002
Trichlorotrifluroethane	F002
White Spirits	D001

        Since most air pollutants are harmful they are generally also classified as hazardous materials with negative utility associated with storage and disposal. Recovery of hazardous materials such as volatile organic solvents for reuse is an excellent example of converting high negative resource utility into significant positive utility—avoiding the high costs of both disposal and procurement. Appendix A lists a number of references that provide case studies and examples of solvent reclamation.

2.2.2   Water Pollution

Wastewater is a very common by-product of many manufacturing processes, primarily due to the abundance and versatility of this undervalued resource. Water is used for cooling in metal cutting, in heat treatment and foundry quenching, as a solvent for rinsing, cleaning, plating, and pickling, and for general housekeeping.

As with gaseous wastes, waterborne wastes are generally considered pollutants and are regulated. Many wastewater sources generate large volumes of water with low contaminant concentrations and are often classified as toxic until treated. Discharge of wastewater is controlled by formal regulation procedures that require the issue of a site permit that specifically identifies the required pretreatment of discharges into waterways or municipal sewers. Wastewater is regulated under one of three general classifications; *conventional, nonconventional,* and *toxic.* Discharge of toxic pollutants is strictly controlled by regulations that require treatment methods based on the *best available technology* (BAT). This term refers to the adaptive nature of current regulations, which require that the most advanced and economically feasible technology available be used for pollution minimization.

Wastewater pretreatment generally requires some method of *waste consolidation* where relatively clean water is extracted to leave a concentrated residue. This residue is then often disposed of in a landfill, and is considered "solid" waste although it is usually in liquid or semiliquid form. It is estimated that as much as 70% of the reported "solid" waste is actually some form of treated or untreated wastewater.

In many cases toxic wastes from sources such as plating solutions and acid baths can be successfully reprocessed to yield chemicals with significant positive utility. Appendix A lists references to examples and detailed case studies of toxic waste reclamation.

2.2.3   Land Pollution

"Solid" waste is so named from regulations that control the interment of waste into solid waste landfills, and, as mentioned earlier, may be misleading regarding specific waste composition. These wastes are further classified by regulation into two categories; *hazardous*, and *nonhazardous*. Solid waste regulations are new to industry and will continue to develop due in part to the rapidly shrinking capacity of available landfill space. Hazardous wastes are strictly controlled by regulations based on the *cradle to grave philosophy*, which saddles the generator of any hazardous waste with the ultimate responsibility of safe disposal. These regulations were a direct response to the environmental damage caused by countless irresponsible and now infamous chemical dumpings known as *Superfund sites,* which created public shock and outrage in the early 1980s.

Nonhazardous Solid Waste. Nonhazardous solid waste (NHSW) is generated from numerous manufacturing processes in such forms as metal cutting scrap, rejected stock material, rejected product, and end-of-life product disposal. This category contains the largest percentage of all waste. It includes industrial as well as municipal waste, and is the only category that does not generally define waste as pollution. Regulation of NHSW is currently based on guidelines that strongly encourage methods for minimizing the creation of waste at the source—otherwise termed *source reduction*.

Methods for source reduction are numerous. Improved material planning is used to avoid or minimize the generation of scrap. Improved production and quality controls help to avoid rejected product. Adopting material tracking and accounting systems can detect inefficient usages. Better training of the work force is another method of source reduction. These are some of the practices that have developed as prudent strategies for improving production efficiency but only recently have been recognized for their environmental value. Other methods for waste minimization focus on materials recovery. These include the recycling of scrap and used materials, and the refurbishment or remanufacture of used components. These strategies are useful but less effective than those already listed, due to the time and energy required for reprocessing. These issues will be discussed more specifically in later chapters of this book.

Hazardous Solid Waste. Hazardous solid wastes are a common by-product of manufacturing processes. This is a consequence of the fact that many useful materials are in some way hazardous. Examples of these include fuels, cleaners, adhesives, sealants, inks, and certain metals and alloys. It is important to note that environmental regulations specifically relate only to the *disposal* of these materials in significant quantities. Their use and their inclusion in consumer products in small quantities are otherwise unregulated. The risk of product liability is at this time the primary motive for limitations in their use by manufacturers.

Hazardous wastes are designated either by *listing* or by *characterization*. EPA regulations maintain four lists with over 400 specific chemical compositions designated as hazardous. In addition, a waste can be so classified if it exhibits any of four hazardous characteristics. These are *flammability, reactivity, corrosivity,* and *toxicity:*

(I) *Flammable wastes* are so classified due to their potential for fire hazard.
(ii) *Reactive wastes* are those which when in contact with other specific chemicals cause dangerous reactions such as the release of toxic gasses. Strict rules prohibit intermingled storage and transportation for this reason.
(iii) *Corrosive wastes* are so classified due to their potential for damage to containers that could cause a risk of spillage and injury.
(iv) *Toxic wastes* are those materials that interfere with biological processes and thus have the potential to cause serious injury or death.

In many cases, more than one of these categories applies to a given waste.

Note that the terms *hazardous* and *toxic* are often erroneously used in this context synonymously. Toxicity is the most difficult category to quantify objectively due to the sensitivity of living systems to extremely low concentrations of toxins and the length of time that may transpire before evidence of any injury can be detected.

Hazardous materials, as previously stated, often have high positive utility. When they become waste they acquire high negative utility due to the expense and liability of storage and disposal. There are two strategies for maximizing the resource utility of hazardous wastes. These are:

(I) *Consume all hazardous materials*: Avoid disposal of hazardous materials or otherwise avoid classifying them as "waste." This may require more careful inventory control for materials with limited shelf life (to avoid discarding expired materials) and better process controls to effectively use all material within a container. Process control improvements to achieve this can often be justified since these materials are typically expensive.

(ii) *Exchange or sell hazardous materials*: Convert hazardous waste into salable or exchangeable materials. Typically a negative utility waste has significant value to other organizations that can either use the chemicals contained in the waste or profitably reprocess it for resale. An example of a waste exchange broker and catalog of listed materials is given in reference 11 of this chapter.

Municipal Solid Waste. Municipal solid waste (MSW) represents the last stage in the consumer product life cycle. It is the most visible and in some ways the most problematic form of waste. The quantity of domestically generated MSW is remarkably vast despite the fact that it comprises *less than* 2% of all solid waste in the United States. The average resident discarded over three-quarters of a ton of trash in 1991, and this figure continues to grow. The major problems of MSW result from the conflicting requirements for landfills that are within close proximity to densely populated areas. Alternatives such as remote disposal incur significant transportation expenses and can be politically unacceptable due to regional boundaries and regulations. Local disposal is desirable because of transportation economics but is undesirable because it increases the risk of contaminating valuable groundwater sources. The added impact of local property devaluation and the stringent regulations for future sites has created a serious economic and political burden for municipalities seeking new landfill sites. These problems are expected to intensify. A full third of the existing landfills will reach capacity before the end of the decade, and pending regulations will demand increasingly restrictive requirements for disposal and monitoring at new sites.

Municipalities have responded to these problems with various interim solutions. These solutions include curbside recycling, environmentally acceptable incineration, and changes in packaging made by manufacturers, either voluntarily or in response to local ordinances [11]. Since recycling and incineration seldom cover capital and operating expenses, these measures are considered temporary strategies to "buy back" landfill space that would otherwise be used for disposal.

Approximately two-thirds of MSW is generated from some form of packaging, with paper being the largest contributor. These materials have potential resource utility as cogeneration fuel and as raw feedstock for reprocessed aluminum and plastics. Unfortunately, the cost of separation and recovery (currently a manual labor-intensive process) generally far exceeds the material's values. Better technology for automated identification and recovery could provide more favorable economics. Materials segregation by consumers provides a significant improvement in recycling economics.

Approximately one-quarter of MSW is composed of durable goods such as appliances and furniture that have reached the end of product life. These items have varying degrees of resource utility ranging from low to very significant for components that are suitable for remanufacture. In many cases durable goods contain significant amounts of toxic compounds that pose a risk for groundwater contamination. Batteries and lead-based solder in consumer electronics are a prime example of this. In these cases materials recovery will ultimately be motivated by the economics of contamination liability that will likely be borne by the manufacturer. Avoiding this potential liability may far exceed the value of these recovered materials.

2.3   EXISTING METHODS FOR WASTE HANDLING

Waste disposal methods depend largely on the potential resource utility of the material. This varies by a large degree as materials move through a product life cycle. Waste generated in the early stages of production is generally high in volume and low in resource utility, for example, mining and other raw material extraction processes. As material moves through the production cycle it increasingly becomes the repository of added value that is generally an aggregate of raw material, energy, labor, and other resources. This cycle ends with the disposal of the product itself with potentially high resource utility for specific materials or components.

Disposal costs have a similar slope that increases through the production cycle and becomes most significant with hazardous materials. Both effects progressively increase the benefit of waste minimization in progression with the stages of the product life cycle. Note that this has a cumulative effect. Each pound of wasted material in later stages of production may previously have generated much larger portions of waste during its manufacture. From an overall life cycle standpoint even small waste reductions can therefore accumulate a cascade of waste reduction from earlier stages of production. The current methods for waste handling and the limitations for extracting useable resources from waste will be discussed in the following sections.

2.3.1   Industrial Disposal

Most of the 6.5 billion tons of waste estimated to have been generated by manufacturers in 1985 consists of wastewater disposed into on-site *surface impoundments* maintained by the manufacturers. These are generally large lagoons that require wastewater discharge permits. This figure does not include significant amounts of off-site disposal—including municipal and commercial landfills, which could not be reliably estimated by the EPA. A large portion of these wastes consists of sludge, oil, slag, ash, food processing residues, solvents, metals, and plastics, with a significant quantity (700 million tons) being classified as hazardous. Manufacturers may conduct some recycling or make arrangements with third-party recyclers.

2.3.2   Recycling

Water, solvents, organic chemicals, plastics, and primary metals are currently recycled in varying degrees by industrial waste generators and third-party recyclers that can sell the recovered materials for profit. The rate of recycling depends on the resource utility of the recovered materials, the technical ease of recycling, handling and transportation costs, and the fear of future liability associated with disposing of reprocessed by-products. In some cases regulations create complications that discourage effective recycling.

Curbside collection is currently the most visible form of materials recycling. Six commodity plastics currently account for 97% of all plastics packaging. At present there exists a viable market for aluminum beverage cans, polyethylene terephthalate (PET), high-density polyethylene (HDPE), and in some cases polyvinyl chloride (PVC). Plastics present the biggest challenge for effective recycling efforts. The advantages for their use in packaging and durable products are numerous, as are the complications associated with recovery and reprocessing. Commingled plastics have very limited market value, and high-quality separation is labor-intensive and expensive. Small quantities of contaminants such as label adhesives and cleaning agents can cause serious degradation that does not become evident until after final extrusion in the next generation product. Even a single bottle of the wrong resin or color can ruin an entire batch of separated plastic. Reprocessing generally results in a resin of lower grade that cannot be used in higher quality, that is, "food grade" products, although methods do exist for producing essentially pure polymers from reclaimed plastic [12,13]. The topic of plastics will be covered in greater detail later in Chapter 6.

2.3.3   Waste Exchange

Waste exchanges provide a service for matching waste generators with waste users. Usually the waste generator does not have sufficient capital or expertise or generate sufficient quantities of waste for efficient materials recovery. Likewise, material users can profitably accept or purchase waste materials and reprocess them in quantity at less expense than buying finished grade raw material. Waste exchanges act as listing brokers for a percentage fee and will find consumers or suppliers of needed materials: usually about 20% of the transacted cost. This can easily result in significant savings in disposal costs for waste generators and raw material costs for users.

This type of third-party business has only recently developed, primarily in response to hazardous waste regulations that have stimulated the development of technology for recovering valuable materials from wastes once thought useless. Other similar organizations will likely emerge in response to other new developments. These will likely be in the areas of recycling and remanufacturing.

2.3.4   Remanufacturing

Remanufacturing provides a means for recycling materials in their "highest value" recovered state. Many typical machine components reach end of life as a result of degradation at relatively few specific locations. This generally includes wear parts such as bearings, commutator brushes, and hydraulic seals; expendables such as ink or printing toner; or life-limited materials that degrade with temperature or environmental exposure. The remainder of the component is often completely unaffected during this time, and includes process-intensive parts such as a shaft, housing, armature, cylinder bore, etc. This type of device can often be disassembled and refurbished by replacing the worn parts at substantial savings in cost as compared to fabricating and assembling a completely new device. Recycling these types of devices by recovering their materials is less efficient than remanufacturing. This alternative requires additional labor and energy to reconstitute the semifinished materials and process them into their original final form.

The viability of remanufacturing depends on many factors. Some of these factors include design features for efficient disassembly, a mature product with stable demand for replacement components for several years, and a positive perception by the public for remanufactured products [14]. Third-party operators have been remanufacturing replacement auto parts such as alternators, water pumps, brake cylinders, and entire engines for over 20 years. Other examples include toner and ink cartridges for printers, and memory chips from computer boards. This topic will be discussed in greater detail in Chapters 5 and 7 of this book.

2.4   CURRENT LIMITATIONS TO ENVIRONMENTALLY CONSCIOUS ENGINEERING

Lack of awareness is the primary limitation to effective environmentally conscious engineering. Design and manufacturing personnel often lack the information resources that could otherwise reveal potential sources of waste early in the product design stage. For example, materials that create expensive disposal problems can be replaced with more environmentally benign substitutes when design personnel are aware of these problems. Knowledge of hazardous waste and disposal regulations will be necessary for effective decision making in these areas. This provides an additional burden to the already complex duties of effective design engineering. Newer simulation and computer-aided design (CAD) tools will be needed for accurate modeling of manufacturing processes with specific emphasis on tracking material yields, wasted material, and disposal costs.

2.5   SUMMARY

Industry generates an extraordinary amount of waste in the course of producing useful products. New requirements imposed by relatively recent environmental regulations have encouraged a review of these practices and have stimulated the development of more efficient technologies. This is especially true for the recovery of waste materials that can return positive resource utility. Wasteful processes are becoming obsolete, and this will eventually stimulate more competitive production methods. In turn, this will favor highly efficient operating procedures that minimize the generation of waste and maximize the recovery of resource utility in wastes generation.

Current practices for recycling, remanufacturing, and waste reprocessing form the basis for future environmentally conscious engineering methods. Design for the environment (DFE) will be based on a combination of existing methods for improved material tracking and quality control, new methods to model manufacturing processes with emphasis on accounting for waste generation, and new methods for more efficient recycling and remanufacturing.

2.6 BIBLIOGRAPHY

1. Francis, B.M. *Toxic Substances in the Environment*, John Wiley & Sons, New York, 1994.
2. *The 1994 Information Please Environmental Almanac*, World Resources Institute, Houghton Mifflin, Boston, 1993.
3. *Industry, Technology, and the Environment: Competitive Challenges and Business Opportunities*, OTA-ITE-586, U.S. Congress, Office of Technology Assessment, Washington, DC, 1994.
4. *Managing Industrial Solid Wastes from Manufacturing, Mining, Oil, and Gas Production, and Utility Coal Combustion*, OTA-O-BP-82, U.S. Congress, Office of Technology Assessment, Washington, DC, 1992.
5. *Industrial Energy Efficiency*, OTA-E-560, U.S. Congress, Office of Technology Assessment, Washington, DC, 1993.
6. *Energy Efficiency: Challenges and Opportunities for Electric Utilities*, OTA-E-561, U.S. Congress, Office of Technology Assessment, Washington, DC, 1993.
7. U.S. Environmental Protection Agency, *Characterization of Municipal Solid Waste in the United States: 1990 Update*, (EPA 530-SW-90-042, NTIS PB90-215112), U.S. Government Printing Office, Washington, DC, June 1990.
8. Kriet, F. *Waste Management: The Engineering Challenge of the 90s*, National Conference of State Legislatures, Denver, CO, 1991.
9. Rechanneling the Waste Stream, *Mechanical Engineering*, August 1989.
10. Puttre, M. Environmental Modeling Helps Clear the Air, *Mechanical Engineering*, pp. 44–51, January 1994.
11. Ehrig, R.J. (Ed.). *Plastics Recycling: Product and Process*, Oxford University Press, New York, 1992.
12. McCrum, N.G., Buckley, C.P., and Bucknall, C.B. *Principles of Polymer Engineering*, Oxford University Press, New York, 1992.
13. James, L. T. *Thermoforming*, Hanser, Munich, 1987.
14. *Design for Manufacturability 1994: An Environment for Improving Design and Designing to Improve our Environment*, Presented at the ASME Design for Manufacturability Conference/ National Design Engineering Conference, DE-Vol.-67, Chicago, March 14–17, 1994.
15. Anonymous. "Question and Answer Sheet," National Corn Growers Association, 1000 Executive Parkway, St. Louis, MO, March 1989.
16. Watson, E.B. "Degradation and Disposal of Packing Material," State Government Technical Brief 75-87-MI, ASME, Washington, DC, November 1989.
17. Rahenkamp, K., and Krieth, F. "A Comparison of Disposable and Reusable

Diapers: Economics, Environmental Impacts and Legislation Actions," National Conference of State Legislatures, Denver, CO, April 1990.

18. Today's Trash, Tomorrow's Fuel, *Mechanical Engineering*, January 1993.
19. Valenti, M. Emission Control: Retrofitting Process Plants for the 1990s, *Mechanical Engineering*, pp. 68–74, June 1992.
20. O'Connor, L. Improving Medical Waste Disposal, *Mechanical Engineering*, pp. 56–59, May 1994.

2.7   PROBLEMS / CASE STUDIES

2.1   Degradable Plastics: Plastics are found in many products produced in America today. As a result, a considerable amount of plastic waste is produced annually in the United States. In 1988, 14.3 million tons of plastic were disposed in solid municipal waste landfills totaling about 9.2% of the waste stream by weight and about 19.9% by volume [7]. According to an EPA projection, the amount of plastic waste could increase to 25.7 million tons by the year 2010. One of plastic's most interesting properties is its chemical stability, which gives it a long life span. Unlike most other waste products, many plastics take hundreds of years to decompose. Consequently, the accumulating amount of plastic litter discarded into landfills takes many years to be reduced by decomposition.

A popular key being developed to help correct the widespread litter problem is the use of "degradable plastics." Many products can be manufactured so that they are either biodegradable (are naturally decomposed by bacteria in the soil) or photodegradable (are chemically decomposed when exposed to sunlight). The obvious advantage to degradable plastics is that it would reduce the amount of time needed to decompose exponentially.

The most available biodegradable plastics are produced by combining cornstarch with plastic polymers. Only about 6% of the composition is cornstarch, which is the only product that will biodegrade [15]. The cornstarch, which is the bonding material, disintegrates, leaving a fine polymer dust.

Questions and problems still exist with the use of these hybrid plastics. The toxicity of the by-products produced from the degradation process is not fully understood.   Additives that make the polymers biodegradable also make it unacceptable for food and retail packaging. In its current state, degradable plastics can only be used selectively. Progress in degradable plastics is still being made so that its uses will have a wide scope and its safety better understood.

A popular application for degradable plastics is in the grocery bag market. Manufacturers have developed a polyethylene bag that has an additive that causes a chemical reaction that breaks down the polymer chains when exposed to sunlight [16]. The degradation continues when the bag is buried underground, where it reduces to an organic powder.

The use of degradable plastics has been mandated by legislatures in some states. Some states have required the purchase of degradable plastics whenever economically feasible to encourage its use [17].

a. Discuss other applications of degradable plastics.
b. Discuss other ways to reduce the amount of plastic waste disposed in landfills.

2.2 Design and Manufacturing for Recycling: Mr. Wiley of Wiley's Car Company is an environmentally conscious manufacturer. He wants to do everything he can to help improve the world in which his grandchildren are growing up. Mr. Wiley has decided that the best thing he can do for the environment is to make his cars as recyclable as possible. In today's economy keeping down costs is very important. Thus, Mr. Wiley needs to make his cars easy to recycle. The goal is to make as much of the car recyclable as possible. There are many methods to make recycling easier. Some methods deal with the design phase of a part and others deal with the manufacturing and assembly of a part. Suggest at least three different techniques for design, manufacturing, and analysis methods to help Mr. Wiley increase the amount of recycling of his cars.

2.3 Energy from Waste: One way to manage today's excessive municipal solid waste (MSW) output is to incinerate it. Waste incineration has two benefits. One is that landfills won't fill and close as quickly, and the other is that energy can be produced by burning the waste. It is estimated that up to 90% (by volume) of waste can be kept out of landfills by incinerating it [18]. Waste incinerators have been around for a long time but were phased out in the 1970s due to pollution concerns and other effects on the environment. Today's waste-to-energy facilities are making a comeback. In 1983 there were just 50 facilities but as of January 1993 there were 142 [18]. The comeback is due in large part to technological improvements that reduce the amount of pollution.

a. Describe how can waste be used to generate energy and discuss the current ways of pollution control for landfills. More specifically, discuss at least two burning operations of waste and describe how energy is obtained from burning municipal solid waste. Can the ash left behind the burning operations be used or should it be disposed of in a landfill?
b. Provide a case study where a local or national industry is using waste to generate energy. Discuss some applications that show the use of this energy.
c. Suppose an incineration facility costs your city $50 million to build. The facility would be capable of burning 650 tons of waste per day producing 13 megawatts (13/24 MW hr) of electricity, which can be sold for $.10 per kilowatt hr. It will cost nothing to

take the waste to the facility, as opposed to paying $50 per ton to have it shipped to a landfill. How long will it take for the waste-burning facility to pay for itself? Ignore operating costs. Assume the facility operates 24 hours per day, 365 days per year.

2.4   Airborne Waste: Recent updates to air-emissions standards have sought to clean up the waste stream flowing into the atmosphere from industrial processes. The Clean Air Act Amendment [19] is requiring the regulation of 189 toxic emissions to the air by industry. To comply with these stringent standards, engineers have developed ingenious new methods of cleaning up and reducing the waste stream before it reaches the environment. In some cases, waste is being stored and processed into useful resources rather than being discarded into the atmosphere. In addition, the dispersion in the atmosphere of the industrial waste stream is being modeled numerically to determine potential environmental impact.

a.   Discuss and list some of the chemicals that are still in use and could be a source of pollution. Consider for example some of the nitrogen compounds exiting industrial smokestacks.

b.   Are there ways to control pollution of these chemicals? Consider for example refrigeration, thermal oxidation, storage underground or in waste ponds on the surface, etc.

2.5   Medical Waste Management: The health care industry in the United States generates roughly 460,000 tons of medical waste annually, and this amount has been growing recently due to efforts to prevent the spread of infectious diseases [20]. In the past, medical wastes such as contaminated paper and glass have been disposed of through incineration or being buried in a landfill. New toxic air emissions standards and regulations on landfill development have led engineers to investigate alternative methods of dealing with the increasing volumes of medical waste. These new disposal methods place an emphasis on reducing waste volume and keeping emissions from combustion very low or eliminating them all together.

a.   Discuss possible environmentally friendly ways of medical waste containment and disposal. Can medical waste be destroyed with "zero" air emissions? How?

b.   Is recycling a possibility in the medical field? Is it a dangerous concept? If not, how about infectious diseases?

APPENDIX 2A: CASE STUDIES AND EXAMPLES OF
HAZARDOUS MATERIALS RECLAMATION

1. International Journal of Environmentally Conscious Design & Manufacturing, Vol. 1, No. 1, Vol. 2, No. 4, ECM Press, Albuquerque, NM, ISSN 1062-6832, 1993.

2. Listings Catalog, Issue 51 (Quarterly Publication), Northeast Industrial Waste Exchange, Inc., 620 Erie Blvd. West, Suite 211, Syracuse, NY, (3154) 422-6572, Summer 1993.

3. Feasibility Study: Pollution Prevention in the Connecticut Metalworking Industry, performed by New England Community Environmental Education Project, June 1990, ConnTAP: Connecticut Technical Assistance Program, Connecticut Hazardous Waste Management Service, Hartford, CT.

4. Hazardous Waste Minimization Feasibility Study, performed by Action Circuits, Inc., ConnTAP: Connecticut Technical Assistance Program, Connecticut Hazardous Waste Management Service, Hartford, CT, July 1990.

5. Feasibility Study: Ion Exchange and Electrolytic Metal Recovery for Chemical Milling Process, performed by Quality Rolling and Deburring Co., Inc., and E & G Equipment Chemical Co., Inc., ConnTAP: Connecticut Technical Assistance Program, Connecticut Hazardous Waste Management Service, Hartford, CT, August 1990.

6. Waste Minimization Audit, performed by PACE, Inc., ConnTAP: Connecticut Technical Assistance Program, Connecticut Hazardous Waste Management Service, Hartford, CT, July 1990.

7. Waste Minimization Technologies Review: Three Case Studies, compiled by Connecticut Association of Metal Finishers, ConnTAP: Connecticut Technical Assistance Program, Connecticut Hazardous Waste Management Service, Hartford, CT, June 1990.

8. Alkaline Cleaner Recycling Using Micro-filtration Technology: 1993 Matching Challenge Grant Program, Automatic Plating Company of Bridgeport, CT, ConnTAP: Connecticut Technical Assistance Program, Connecticut Hazardous Waste Management Service, Hartford, CT, 1993.

9. ConnTAP Quarterly (quarterly newsletter), ConnTAP: Connecticut Technical Assistance Program, Connecticut Hazardous Waste Management Service, Hartford, CT.

THREE

NEW THEORIES
FOR ENVIRONMENTAL REGULATION

Humankind is the beneficiary of the industrial revolution and the environment has been the victim. Industry has produced items of every kind to aid humans, but unfortunately industry did not take care of the environment. Prior the 1970s, there were no laws to protect the environment from the side effects of the Industrial Revolution, mainly pollution. Many factories produced as much environmentally damaging pollution as they did finished products. The trend over the past two and a half decades has been an increase in government regulations on the amount of pollution or waste that industry can put into the environment. In this chapter, a brief synopsis of the evolution of environmental regulation is laid out to aid in understanding the sensitivity of environmentally safe products. A time frame is established concerning the beginning of public awareness of the detrimental effects of the various forms of industrial and municipal pollution that has been occurring for decades. Legislation is explained with emphasis being placed on the important role of the U.S. Environmental Protection Agency (EPA) and the agency's forms of policies.

3.1   INTRODUCTION

Environmental law as presently known is a relatively recent development. Prior to 1970 there existed few explicit regulations to restrict the discharge of harmful materials into the environment. Before this, the only effective means of restraining polluters was through complex litigation known as the *nuisance claim*. Nuisance claims are derived from a category of *private law* known as *tort law*, which provides a formal means to resolve injury disputes between private parties.

Filing a nuisance claim places the burden of legal proof with the private accuser who also must bear the financial burden of legal action. The accusers must show evidence that they have been harmed and legally prove that the polluter is the cause of the offending nuisance. In environmental cases this created serious problems. The defendant (the polluter) could often avoid culpability if the source was not specifically identifiable, as when multiple polluters discharge similar chemicals into a common waterway. It could be successfully argued that other sources might be the true cause of the claimed nuisance for which the defendant was not responsible. Simply proving harm could also become a contentious procedural battlefield. Commonly the polluter was an entity with greater financial and legal resources than the accuser and could often bear the cost of protracted legal action as a cost of business [1-5].

In theory, a group of polluters could be held accountable by a claim of *public nuisance*, although this proved equally ineffective. Public nuisance claims require an elected official to take legal action against a group of offending parties. For practical reasons public officials are often reluctant to act against such groups, which generally possess considerable political influence and often provide employment to a significant fraction of their voting constituents.

Effective pollution control was therefore largely a matter of conscience. Decisions to limit the discharge of harmful materials were made by operators on the basis of common practice and the legal willpower of their neighbors. Unscrupulous operators who placed little value on environmental quality chose to ignore the need to minimize pollution and set the unwritten performance basis for other businesses. In this sense, such businesses *externalize the cost* of maintaining a clean environment to the surrounding community. Over time their competitors found little benefit in not following suit, since perception became to be, *the immediate benefit to the individual overwhelms the intangible negative consequences to the community.* This syndrome has been known from antiquity as the "Tragedy of the Commons" [6].

In 1969 the Cuyahoga River near Cleveland, Ohio, caught fire due to tremendous concentrations of volatile chemicals. That same year a large oil spill off the coast of Santa Barbara, California, stirred a public outcry for corporate and industrial accountability. In most major cities rivers were unfit for fishing or swimming and smog levels became a routine element of weather reporting. The absorptive capacities of the air and waterways had reached their limit. Clearly the maintenance of environmental quality through the mechanism of tort law was not working.

The decade of the 1970s marked the beginning of our present system of environmental law: founded in *public law*. In 1970 the U.S. Environmental Protection Agency (EPA) was formed as a consolidation of 15 departments from 5 separate federal agencies. The EPA was chartered to define systems of enforceable rules as a more effective means to restrict industry and the public from polluting the environment. Through this rule, many activities became publicly subject to explicit limits and no longer required the action of private litigation. This included such things as emissions limits for fossil-fueled power plants or the need for specific types of technology, such as catalytic converters for automobile manufacturers. The size of this agency and other similar agencies, the volume of regulations, and the scope of environmental problems identified since its inception has grown dramatically during this latest generation. This is due in part to the disastrous neglect of prior years.

The EPA was intended to consolidate environmental policy and lawmaking. Unfortunately, it never actually achieved this goal. The EPA is now only one of many agencies responsible for establishing environmental laws.   Many other federal agencies as well as state and local authorities now impose restrictions on public and business activities that adversely affect the environment.   The complexity of these sometimes conflicting laws has become a serious concern for businesses and the agencies responsible for enforcement [7]. These concerns have recently prompted a number of agencies to coordinate their enforcement activities and regulations to provide more simplified and less time-consuming interpretation by businesses.

A comprehensive guide to all regulations is beyond the scope of this text. In this chapter we are primarily concerned with the fundamental basis for environmental regulation with specific emphasis on trends in legislation. These newly emerging philosophies, regardless of just how soon they are legislated, will ultimately establish the requirements for environmentally conscious products and processes in the future.

3.2   LEGISLATION

Environmental laws begin with the passage of legislation that gives specific authority to an *administrative agency*. When a new agency is created this law is termed *enabling legislation*.[1] The Federal Communications Commission (FCC) and the Nuclear Regulatory Commission (NRC) are just two examples of the many other administrative agencies. Administrative agencies are responsible for studying specific issues, considering various regulatory options and policy implications, and establishing proposed regulations into law. For the EPA these laws are then *promulgated* to the states for enforcement. This means that the states are given a limited time to officially accept and integrate them as the minimum standards for local laws. These generally require the states to provide their own operating revenue (often through taxation), which has been known to create conflicts between state and federal authorities, as well as marginal degrees of enforcement by some reluctant states. Conversely, some environmentally proactive states (such as California and Oregon) exercise their right to impose more restrictive regulations than the minimum standards promulgated by the EPA. This is only one of many complications that must be addressed by manufacturers who sell products and must comply with local regulations throughout the country.

Administrative agencies can receive delegated authority from any of the three branches of government; the EPA receives authority from all three. It conducts hearings and writes laws (legislative power), it forms committees to investigate wrongdoings (executive power), and it adjudicates disputes with businesses by an internal court system (judicial).   Some have criticized this present form of administrative agency as a dangerous consolidation of powers specifically separated by the Constitution and which has thus grown into an unofficial "fourth branch" of

---

1. The EPA was actually established without the action of congress through an executive order by then President Nixon.

government. Although powerful, administrative agencies remain under the control of their delegating branches of government and must operate within a framework defined by their enabling charter. They therefore have less autonomy than is often claimed.

3.3   ENVIRONMENTAL POLICIES

The debate continues on many issues about what constitutes justifiable regulation. Businesses have claimed many rules to be unnecessary and expensive, while the public grows more doubtful of air and water quality. These issues are complex, and fair resolution begins with a definition of priorities and goals. Fortunately, the philosophical basis for environmental policy has expanded in the short time since formation of the EPA and now guides the development of newer regulations. In the most general sense an environmental philosophy can be categorized within a broad spectrum of five generally accepted paradigms. This spectrum provides some context for understanding the issues that are fundamental to debates of environmental policy and the future development of regulations. The spectrum is described as follows.

3.3.1   Paradigm 1: Frontier Ecology

This philosophy represents an absolute lack of environmental protection and generally occurs where only free markets rule. There is no mechanism for maintaining environmental quality. These costs are therefore externalized at every opportunity. Even the most exploitative domestic businesses recognize this attitude as unacceptable for the developed world, although much of the third world still operates at this level and the results are obvious and devastating.

3.3.2   Paradigm 2: Environmental Protection

This includes the initial stage of public environmental authority and approximates the current stage of environmental policy in this country. Authority is derived from the tradeoffs between the cost to the individual and a minimum level of public health and safety. This includes some degree of concern for other forms of life such as endangered species. The most common mechanism for exercising control is through *command-and-control* or *end-of-pipe* regulations. These mandate compliance with specific standards through threat of fines and possible criminal penalties. Most regulations focus on controlling pollutants in the final stages of manufacture and sometimes impede the development of newer, less intrinsically polluting processes due to the inflexible and imperative nature of the regulations. This often leads to unproductive adversarial relationships between regulators and businesses that place strategic emphasis (and large financial commitments) on litigation.

3.3.3   Paradigm 3: Resource Management

In this view the earth is recognized as a closed *economic* system. If the true costs of environmental maintenance can be accurately identified, quantified, and *internally distributed*, then the free market will encourage the development of technology for the most efficient use of resources and will avoid the pollution-generating processes. Authority is derived by determining economic equivalents for specific levels of *ecologically sustainable* pollution and nonrenewable resource consumption. New mechanisms can then be used for more efficient environmental controls such as *green taxes* and *tradeable pollution permits* [8–11]. Green taxes raise revenue from the use of virgin raw materials, thus stimulating the development of materials recycling technology. Tradeable pollution permits offer a polluter the option of selling its permits at a profit, if the polluter can reduce its pollution levels by any other means with an equivalent or lower cost. This provides a profit incentive to develop cheaper methods for pollution reduction by any means available without the intervention of cumbersome regulations.

Some existing regulations have progressed to this stage. A system of tradeable air pollution permits now exists for power generation plants in the United States, and ozone-depleting chlorofluorocarbons (CFCs) are subject to increasing green taxes. The problem with these methods of regulation lies in the difficulty in establishing monetary values on intangible qualities such as pollution levels.

3.3.4   Paradigm 4: Eco-Development

This paradigm raises the value of the ecosystem to that of human society and represents a fundamental moral and ethical shift in societal attitudes. This departs from the traditional view held by modern mankind that survival demands the conquest and exploitation of nature. The human economic system here would become a subordinate element of the global ecology. Policy objectives within this paradigm would focus on a closed-loop material cycle and minimal consumption of nonrenewable resources. Regulation is not expected to progress to this stage soon, although it does provide a theoretical model for sustaining the earth and humanity in perpetuity. The problem with this model is the basic incompatibility of expanding human population and economic growth with the perpetual survival of the planet.

3.3.5   Paradigm 5: Deep Ecology

This paradigm represents the deepest level of personal commitment to the ecosphere and might be characterized by identification with the spiritual myths of primitive humans, or the desire to "return to nature" as a state of pure harmony. We refer to this paradigm here because it serves as an inspiration and possible idealized life-style to many people. Like the first paradigm it serves as a marking post and does not provide a rational basis for environmental regulation.

3.4   SUMMARY

The area of environmental law was not well defined prior to the 1970 creation of the EPA. No specific definitions of what pollution is were in use, so an accuser had to prove that some harm was done to the environment to be compensated by a polluter. It was burdensome for the accuser to make a case against a single polluter, because there could be many sources of the pollution. In addition, stopping one source of pollution did not protect the environment from other sources of pollution.

The creation of the EPA created a set of standards that industry is held to. Under the laws of the EPA, all polluters are held to the same restrictions on emissions. Although the creation of the EPA made regulating polluters less difficult, some members of industry claim its laws are too complicated and restrictive. This is complicated by the fact that state and local agencies often have their own set of environmental standards that sometimes contradict those of the EPA.

3.5   BIBLIOGRAPHY

1.   *In the Age of the Smart Machine: The Future of Work and Power*, Basic Books, New York, 1989.

2.   *Understanding the Small Quantity Generator Rules: A Handbook for Small Business*, U.S. Environmental Protection Agency, Office of Waste and Emergency Response, EPA-530-SW-86-019, Washington, DC, September 1986.

3.   *Small Quantity Generator Guidance for Hazardous Waste Handlers*, State of Connecticut Department of Environmental Protection, Bureau of Waste Management, Hartford, CT, March 1993.

4.   *Pollution Prevention Options Fact Sheets for Industry*, State of Connecticut Department of Environmental Protection, Bureau of Waste Management, Hartford, CT, June 1993.

5.   *Hazardous Waste Management Regulations* 22a-449(c)-100 through 110 and 22a-449(a)-11, Regulations of Connecticut Agencies, State of Connecticut Department of Environmental Protection, Bureau of Waste Management, Hartford, CT, revised July 1990.

6.   Hardin, G. The Tragedy of the Commons, *Science*, 168, 1968.

7.   The Missing Link in ECM: An Intelligent Advisory System for Regulation Interpretation, *International of Journal of Environmentally Conscious Design & Manufacturing*, 2(1), pp. 31–38, ECM Press, Albuquerque, NM, 1993.

8.   *Green Products By Design: Choices for a Cleaner Environment*, OTA-E541, U.S. Congress, Office of Technology Assessment, Washington, DC, 1992.

9.   *The United States Government Manual*, Office of the Federal Register, Washington, DC, 1992.

10.   Kubasek, N., and Silverman, G.S. *Environmental Law*, Prentice Hall, Englewood Cliffs, NJ, 1994, ISBN 0-13-285107-5.

11.   Hawken, P. *The Ecology of Commerce*, Harper Collins, New York, 1993.

12.   Kriet, F. *Waste Management: The Engineering Challenge of the 90s*, National Conference of State Legislatures, Denver, CO, 1991.

13.   Malaspina, F. "Cycle Life of Communications Batteries and the Impact on Waste Reduction," Proceedings of the 10th Annual Battery Conference on Applications and Advances, 1995.

14.   Kiehne, H.A. "Batteries and Environmental Requirements," 13th International Telecommunications Energy Conference, 1991.

15.   Phillips, J. Power Sources and Portable Computers, *IEEE Aerospace and Electronics Systems Magazine*, May 1993.

16.   Hendrickson, C. Product Disposal and Re-use Issues for Portable Computer Design, *Proceedings of the IEEE International Symposium on Electronics and the Environment*, 1994.

3.6   PROBLEMS / CASE STUDIES

3.1   Make a detailed study of products and processes that use hazardous materials that cannot be replaced, per se, such as alkaline and nickel cadmium (Ni-Cad) batteries. Evaluate these batteries based on capacity (mA-h), cycle life, life-time power (A-h), life-cycle cell requirements, weight, disposal fee, cost per cell, the cost of the electricity to power the recharger, energy consumed (watts), and the cost of the rechargers themselves, batteries purchase cost, disposal cost, energy cost, life cycle cost, etc. Note that the alkaline batteries contain hazardous materials such as mercury and lead [13]. However, the alkaline batteries manufactured and sold today in Germany contain so little hazardous material that there is no longer a need to dispose them of in special landfills. They can be disposed of as normal household waste [14]. Ni-Cad batteries, on the other hand, contain cadmium, which is a carcinogen and can also cause renal failure, and nickel, which is a carcinogen [15,16]. These batteries must be disposed of in a hazardous waste landfill or recycled. Compare these batteries with the nickel metal hydride (Ni-MH) batteries used in portable computers. Devise reliable ways to prevent these materials from reaching the wrong places—landfills, groundwater, food chain, etc. Are there other ways to do this besides trusting the end user to dispose of the material properly?

3.2   Household appliances waste energy because they are left on when people are sleeping or out of the room. Often television sets, room lights, radios, and room fans are left on when no one is around to use them. Commercially available devices such as the "Clapper" are some means of conveniently turning on or off most of household appliances. The Clapper device allows the user to remain where he or she is and turn appliances on or off by clapping hands. But the question remains: What happens if the person forgets or falls asleep? A new device is needed to address this problem. Through your understanding of the five paradigms discussed in Section 3.3, propose a system to help save this energy. Hint: You might look into the design of a new motion detector that automatically turns off many common appliances that are not useful when people are not around to enjoy them.

The detector should be able to shut off the appliance if it does not detect motion for a period of time. Also study the effect of not moving for periods longer than the programmed one or having the feature of deactivating the detector if not needed.

3.3   Metalworking operations including cutting, stamping, deburring/tumbling, heat treating, phosphate plating, and degreasing require the use of specific processes and chemicals that contain or generate potential pollutants. Pollutants can be contained in cutting fluids, degreasing solvents, or in the chemicals of the phosphate-treating or heat-treating processes. Metal chips in these liquids present an added problem to waste disposal procedures.

a.   For each of the metalworking operation just mentioned, discuss the specific pollutants and the problems they can create.

b.   Discuss possible solutions to these problems. For example, cutting operations require cutting fluids (pollutants). Used fluids could be contaminated with metals and require special filtration disposal, etc. (problems). These problems could be handled by improving the currently used disposal techniques, acquiring modern filtration equipment, etc. (solutions).

3.4   Municipal Solid Waste: A major dilemma facing America today is the safe management and disposal of the seemingly endless supply of municipal solid waste (MSW) produced by this country. Conventional practice of disposing of solid waste in landfills has become unacceptable because of the environmental and health hazards it presents. Projections predicting the amount of municipal waste to increase from 160 to 190 million tons per year in this decade, along with the fact that many local landfills are reaching their capacity limits, indicate the critical need for new policies and practices in the disposal of solid waste to be implemented [12]. The EPA has taken the lead in devising new policies for the disposal of solid waste. One of the EPA's approaches to reducing the amount of waste relies on voluntary participation, not mandatory compliance to rules or laws. The EPA's strategy is to provide "information, technology transfer, and technical assistance to industries that generate waste."

a.   Discuss some of EPA's goals in reducing solid waste. What are other ways that the EPA can help to encourage the public's participation in the reduction of MSW?

b.   What are the positive and negative benefits to a system whose success relies on the voluntary participation instead of mandatory enforcement?

3.5   In the software industry, a great deal of waste is generated through the disposal of unsold and returned software diskettes. Instead of forcing a new law to regulate the use of these diskettes, can you recommend a voluntary method that industry can put into practical use? For example, you might suggest the reformat and reuse of unsold diskettes. For the case of defective diskettes, you might suggest an economical and yet practical way of recycling the different plastic and metal parts of the diskettes. Explain and document your solution.

FOUR

DESIGN FOR THE ENVIRONMENT

No engineering/manufacturing methodology has sparked more interest in recent years than design for the environment (DFE). While design for manufacture and assembly (DFMA), concurrent engineering (CE), and design for disassembly (DFD) also capture their share of headlines, with DFE comes the promise of reversing decades of neglect regarding the environmental responsibility of designers and manufacturers. In addition, DFE provides a good tool for conserving and reusing the earth's scarce resources; in a sense, our planet can indeed be preserved for future generations. Engineers really have no choice: by sheer necessity, DFE will soon represent the basis of all their future efforts. However, DFE presents unique challenges to designers and manufacturers alike. The biggest and broadest of these changes is the lack of an all-encompassing "industrial ecosystem", where energy and material consumption are optimized, minimal waste is generated, and output waste streams from any process serve as the raw materials (inputs) of another. Therefore, this chapter will examine why designers and manufacturers need to change their "modus operandi" in the first place. Next, it will derive general DFE guidelines and their ramifications for industry and manufacturing. Finally, success stories from industry and specific DFE applications used by the design and manufacturing community at large will be addressed, with particular attention placed on the use of the derived guidelines.

4.1   INTRODUCTION

Given the wide reporting of environmental concerns today, it's very difficult not to be pessimistic about the future of our planet. From the days of the Industrial Revolution, technology has spurred civilization on to new and staggering heights in very short spans of time. New technologies have made our lives much easier, but past designers didn't always stop to consider undesirable side effects. The development of refrigerator compressors suitably illustrates this point: Early compressors used ammonia or sulfur dioxide, both of which are toxic chemicals that sometimes injured and even killed people [1]. Then chlorinated fluorocarbons (CFCs) were developed and were hailed for their safety, low cost, and use in popular applications such as air-conditioning. Only later was CFC use connected to phenomena like global warming and ozone layer destruction.

The "family of man" is always growing. By the year 2030, 10 billion people will probably live on this planet. Critical natural resources will be expended to support this population and enormous waste streams will be generated. Even today, each person in the United States alone produces a daily average of 4.5 pounds of solid waste. The result is approximately 180 million tons of municipal solid waste sent to landfills annually. By 2030, that figure could climb to 400 billion tons annually, which is "enough to bury greater Los Angeles 100 meters deep" [1]. With so much taken out and none being put back, it doesn't take much thought to conclude that we are on a collision course with environmental disaster. We must find efficient ways to recycle the waste streams and put them back to work in the ongoing "industrial ecosystem."

In the light of the previous discussion, it's not difficult to see how DFE can contribute toward minimizing the waste generated. Manufacturers produce goods for public consumption; because few mechanisms exist to recycle most products at the end of their useful life, landfills are put into service with the consequences already outlined. Thus, DFE dictates that the designer/engineer should examine products for environmental soundness, which encompasses an environmental "cradle-to-grave" snapshot across the entire product life cycle. The traditional product life cycle, as illustrated in Figure 4.1, must change to ensure that a DFE infrastructure is in place.

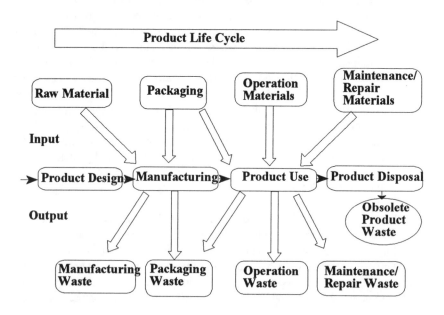

Figure 4.1 Material input and waste streams: Traditional product life cycle.

The motivation for DFE should stem from more serious concerns than just reducing costs and minimizing waste. Fortunately, in many instances, the latter are also realizable, which makes DFE a methodology worth considering for more than purely altruistic reasons.

Further, environmental legislation nationwide and worldwide is more stringent than it has ever been. Federal, state, and local statutes regulate air emissions, water discharges, occupational exposure, and the treatment and disposal of various hazardous chemicals [2]. The 1990 Clean Air Act emphasizes waste minimization as well as emission monitoring and reporting, and lists 189 compounds for regulation as air toxins. In addition, the internationally ratified Montreal Protocol regulates the production and use of halogenated organic compounds (CFCs are among them). With such severe legislative pressures being brought to bear upon manufacturers, the final outcome is clear: Changes in manufacturing and design philosophy must occur.

With this very brief introduction, the stage for a closer look at the DFE process is now set. The "leading-edge" efforts of manufacturers today and highlights of the major challenges that DFE presents will be examined in the following sections.

4.2   DFE EXPLORED

From concurrent engineering practice, it is known that approximately 75% of a product's total life cycle costs is determined in the design stage. Therefore, decisions made during the design phase profoundly impact the entire product's life cycle. This provides the incentive for "doing one's homework" very early on in the design process. To avoid environmental problems after the design phase, there is no apparent reason why environmental concerns could not also be addressed in the initial design stages. The overall costs associated with waste streams can then be reduced [2].

Besides the reduction or elimination of product waste streams from the manufacturing process, the designers must examine the environmental impact of the

Figure 4.2 Material input and recycle streams: Ideal product life cycle.

design when it is being produced and disposed of or recycled as described in Figure 4.2. Design for recyclability (DFR) can play a significant role in this effort. DFR is an infrastructure where products can be accepted at the end of their useful life by other processes. Thorough materials research at the beginning of the design phase will go a long way in improving the recycling process. To enhance eventual breakdown and recycling, design for disassembly (DFD) principles must be used before the design is finalized. The times, techniques, and costs to disassemble a product are just as important as those factors that exist during its assembly. Since time is money, it is obvious that DFE cannot be implemented without added costs.

When environmental regulations become more strict, waste-associated costs increase. In addition, new studies and better chemical analyses can result in the recategorization of waste from nonhazardous to hazardous, "resulting in a tenfold increase in disposal costs" [2]. The obvious solution is to reduce waste in the first place, but more important, to ensure that any kind of waste (to include scrap, material rework, and large unused inventories) is minimized in all operations. Not surprisingly, this is exactly what is propounded by just-in-time (JIT) and concurrent engineering; all production operations should be optimized to prevent waste and thus reduce costs. Sometimes this optimization can involve the use of design for manufacture and assembly (DFMA) principles; if a product takes less time to assemble, costs are further reduced. As recyclability of the product at the end of its life is a concern for the DFE designer, the designer has the added problem of simultaneously designing for disassembly (DFD). Once again, if disassembly time is reduced, so are associated costs. Naturally, any cost saving over the entire product life cycle will benefit both the consumer and the producer.

DFMA and DFD philosophies may appear contradictory, but sensitivity to both aspects can ensure the design is sound for assembly as well as disassembly. For instance, dual-purpose snap fits can eliminate fasteners (a DFMA plus) and facilitate disassembly (a DFD plus) [3]. The Boothroyd and Dewhurst DFMA model may be used to predict times and cost for assembly [4]. Also, recognizing that the actual assembly sequence isn't always the reverse of the disassembly sequence. Arai and Iwata [5] developed a computer-assisted design (CAD) system with a product assembly/disassembly planning function. Based on kinematic simulation, the latter is one example of a planning methodology that can be utilized in the earliest stages of design. Its use in CAD software allows numerous changes and iterations, "what if" analyses, etc., and provides the engineer with a tremendous amount of flexibility in design.

Another example of a DFD software design tool that integrates DFE elements is the Restar, developed by researchers at Carnegie-Mellon's Center for Integrated Manufacturing Decision Systems in Pittsburgh, Pennsylvania [6]. Restar analyzes disassembly operations with regard to time involved, cost and effort of each step, value of recovered parts and materials, and any savings in energy and pollutant emissions. This tool quantifies the results of the DFE process for the designer's benefit, and also shows how design changes may further improve those results. Other software tools covering material selection are in development and compilation of a comprehensive database of the environmental impact of various material choices.

A recent DFD seminar [7] focused on what industry can do upstream to reduce the enormous amounts of plastic that are currently destined for landfills. The outcome of the seminar was three primary recommendations: source reduction, DFD, and use of recycled materials. It was emphasized that source reduction treats the symptom (full landfills), not the cause (our inability to recycle certain materials and products); still, it finds use as a short-term strategy. It was also emphasized that reducing the total number of materials in the design is crucial to DFD. It was observed that the only answer is to overdesign the material performance in some parts in order to match materials with the parts requiring those additional characteristics. In so doing, the designer decreases separation time and increases the product's value to the recycler; costs may increase initially, but will soon be canceled out by increasing disposal costs. Other material selection options include substituting equivalent materials to suit performance specifications or using two compatible materials that can be recycled together. It was stated that "Ideally, the goal of material selection in DFD is to eliminate the need for disassembly" [7].

Specifying and using recycled materials in original designs will create and support the markets for these materials in the future. Unfortunately, many designers refuse to use recycled materials because they don't believe they're as good as virgin materials. Therefore, like GE Plastics of Pittsfield, Massachusetts, even material suppliers involved in recycling put most of their development effort into virgin materials [8]. This perception must change for DFE to succeed. Further, DFE is now seen to be so closely related to other design disciplines that the whole "design-for" realm is increasingly referred to as DFX [9].

The DFE designer must ask and answer a long list of questions about the product and process, from beginning to end of the life cycle. The following questions are only a few that exist among a long list [3, 10]:

- What material properties are desirable? What strength and functional properties must the raw materials and the finished product possess?
- To conserve natural resources, can recycled materials be used instead of virgin raw materials? Will these materials provide the same "fit and function" to the product? If not, do other viable alternatives exist?
- What added costs (capital and operational) will be incurred if the new materials are used with (perhaps) different production processes?Has the production process been examined for waste generation? Are the solvents and fluids in use hazardous to humans or the environment? Are nontoxic substitutes available? Or can these solvents be eliminated from the picture by using a different manufacturing process? What added costs will be incurred?
- Has the product been designed for optimal DFMA and DFD? Are specifying tolerances applied to individual parts? If so, will the completely assembled product meet the required overall tolerances?
- Have individual parts been marked with molded-in logos (to identify material types) and part separation points to facilitate breakdown and recyclability?
- Is a standardized industrial coding and information database available to track recycled material availability as well as breakdown and recycling facilities?

This is more in line with the all-encompassing "industrial ecosystem" and DFE infrastructure discussed earlier. For DFE to work, such a system will be absolutely vital. Fujitsu Ltd of Japan claims its most significant achievement in the domain of environmental planning has been the development of an on-line database for environmental information [11]. It is accessible via the company's personal computer network where data on properties, handling, hazards, and disposal is readily available. While this network may not appear very comprehensive, it certainly represents a step in the right direction.

The preceding list is not all-inclusive, but several points become clear. First, not all the design considerations are necessarily environmental, but, interestingly, they fold in quite nicely with the other "optimal" design and manufacturing methods (DFMA, JIT, CE, etc.) that fall under the umbrella of "continuous improvement." Second, industry must put out a lot of effort to establish a comprehensive nationwide and international DFE infrastructure. Effort equates to time as well as money, so while one can see this is not a trivial task, it is certainly within the realm of possibility. Finally, designers must use their creativity in several different ways to come up with efficient, "green" product designs for future generations. Fortunately, the designers have several tools (such as CAD/CAM and DFMA software) at their disposal.

4.3   DFE PRESSURES

In this era of environmental awareness and (sometimes) activism, several diverse groups have pressured manufacturers to change their ways. Recyclers have tried for years to get manufacturers to use DFR [12]. Legislation has been introduced in this country and abroad to get manufacturers to think in terms of environmental protection and recycling; the Europeans appear to be far ahead of the United States in this regard. In August 1992, Germany's environmental minister Klaus Topfer proposed a rule that "would require all automakers, or their authorized representatives ("authorized representatives" includes car importers—U.S., Japanese, French, British, etc.—and their suppliers), to take their cars back from consumers for recycling. The rule, which started in 1993, also mandates that automakers recycle 20% of plastics by weight in cars starting in 1996 and 50% by 2000" [13]. While the German auto industry already recycles 75% of its cars by weight, more is expected of it so the ramifications are global in scope. In the United States, the House of Representatives has considered amendments to the Solid Waste Disposal Act that would require manufacturers to take back products that cannot be recycled [8]. In 1990, the Energy and Commerce Committee introduced several bills addressing recycling of appliance components (H1810, H2845, H3735) [3]. Finally, consumers themselves have spoken through the use of surveys: 90% of German consumers said they "favored environmental protection measures and preferred clean, green products" [14].

And how have manufacturers and designers responded? Environmentally sound efforts abound in the United States and Europe. For example, Black & Decker of Canada has started a "take back" program for its cordless products and recyclable nickel-cadmium batteries. The consumers return the used products to Black & Decker authorized service centers, where they receive a $5 rebate toward their next purchase [12]. In addition, Kodak's "single-use" camera is recycled after it is returned for film processing. Also, General Electric designed a refrigerator, called the Totem, that can be disassembled and recycled [12]. As will be seen in the following sections, automakers in the United States and abroad and other large producers of consumer goods are incorporating DFE into the design and manufacture of their products and/or doing research to arrive at more reliable, practical, and recyclable plastics and nonhazardous cleaning solvents [8, 15, 16]. The battle continues with new ground gained every day!

4.4   DFE GUIDELINES

Based on the preceding discussion, it is possible to summarize some of the general guidelines used in DFE. Again, an environmentally sound design (the flower in Figure 4.3) can grow in a "green" engineering environment (the sun) where the designer can understand the diverse nature of DFE and deal with it accordingly. In addition, there is no substitute for comprehensive knowledge, solid research, working partnerships, adequate funding, and a firm commitment from upper management.

With this understanding and the ideas presented in this chapter, some not-all-inclusive DFE guidelines may be stated as follows [9, 10]:

1. Keep the design simple by using as few materials as possible. Also, incorporate as many functions as possible into any single part without compromising function. Avoid secondary finishes, toxic materials, and heavy metals that can contaminate the material.

2. Find multiple or secondary uses for a product. Disposal will be less of a problem if a product has more "intrinsic value." As an example, a container protecting a

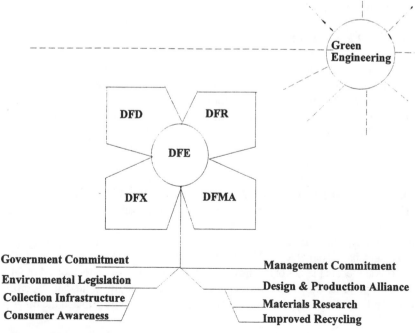

Figure 4.3 The growth of DFE.

product could also store the accessories that come with it.

3. To ensure easier recycling, use materials that match each other closely or are of the same material. Look for ways to use recycled materials as starting compounds for a product.

4. Modular design should be preferred whenever possible. This helps in maintenance and repair—a "black box" concept.

5. Design for long product life and become more service oriented. If a manufacturer upgrades its product as technologies improve, a more loyal customer base is assured.

6. Ensure tracking mechanisms are available on a "cradle-to-grave" basis. Ensure up-to-date databases are available. Ensure parts are marked with logos to aid in recycling efforts.

7. Examine those components in a design that may be reused upon failure or disassembly. This would reduce the need to recycle or dispose.

8. Establish a network of producers and suppliers to form the beginnings of the "industrial ecosystem" and facilitate DFE efforts.

9. Look to reduce waste by-product streams in manufacturing processes. Seek out nonhazardous solvents and cleaning materials. Reduce energy consumption by eliminating unnecessary manufacturing steps.

10. Ensure a product "buy-back" infrastructure is in place and well advertised to suppliers, producers, and consumers.

11. Pay close attention to recyclability and reuse of packaging, shipping, and other peripheral requirements. Design reusable shipping vehicles.

12. Whenever possible, attempt to incorporate a concurrent engineering philosophy—JIT, DFMA, DFD—to aid the overall DFE effort. Design for total ease of assembly, separation, handling, and cleaning.

13. Apply tight tolerance design principles to reduce the use of fasteners and keep the separation process simple.

4.5   DFE APPLIED

As stated several times during the course of this chapter, DFE principles are being applied very successfully today by manufacturers in diverse industries. It is their creative efforts that provide optimism for the future of our planet, both from a social and from an industrial perspective. The following industrial applications examine specific DFE efforts.

4.5.1   The U-Kettle

The U-Kettle was designed by Polymer Solutions of Worthington, Ohio, for Great British Kettles, a United Kingdom-based appliance manufacturer [3]. Polymer Solutions is a joint venture between GE Plastics and Fitch Richardson Smith, an industrial design firm. Besides a metal heating element, the electric U-Kettle consists of six parts injection molded of Noryl modified polyphenylene oxide (PPO) supplied by General Electric Plastics of Pittsfield, Massachusetts, and two parts molded of GE's Lomod copolyester elastomer. The six parts are the reservoir, base, lid, handle, cross-brace, and on/off button, while the last two parts are the handle grip and the lid grip.

Mack Molding of Arlington, Vermont, is the appliance molder; an initial run of 500,000 units required about 400,000 pounds of plastics, since each kettle consists of 21.5 ounces of Noryl and 2.6 ounces of Lomod.

In this design, dual-purpose snap-fits were used to eliminate fasteners and enhance disassembly at the same time—a good example of DFMA and DFD principles used in concert. Molded-in break points were also designed in without any loss of performance during normal operation. Another area of concern during product prototyping was the handle of the assembled kettle, which proved to be unsteady when filled; sonic welding of two points took care of the problem without affecting disassembly. Finally, to aid recycling, molded-in logos clearly identified material type. To aid disassembly, molded-in instructions clearly showed part separation points.

The U-Kettle is a classic example of what a DFE effort should represent. The partnership between an industrial design firm, a plastics producer, and an appliance manufacturer effectively married three critical areas of DFE: design, material selection/properties, and the manufacturing process. The "experts" in each area were able to put their heads together to arrive at a producible, environmentally sound design. They recognized the lack of and need to push for a recycling infrastructure and ensured both thermoplastics used could be recycled together. Also, the U-Kettle is reported to consume less electricity than any other electric cookware item, which is another victory in the fight to conserve resources [9]. Finally, the designers and manufacturers worked together using DFMA, DFD, and DFR creative design solutions to overcome practical problems. Their "trial-and-error" solutions have blazed a trail for future DFE designers and manufacturers.

4.5.2 Opel's DFE Program

Even though a lot remains to be done, several shining DFE successes in the auto industry stand out. The German manufacturing arm of General Motors (GM), Adam Opel AG [17], plans to use 22 million pounds per year of recycled plastics in more than 20 automotive parts. Furthermore, the company plans to recycle nearly 100% of the plastics used in its new models, called Astra and Calibre. Opel expects plastics to comprise about 15% of a car's weight by the year 2000, up from 10% in 1985. Fuel efficiency should improve through light-weight components and aerodynamics. Also, Opel made a conscious DFE decision not to use tough-to-recycle reinforced sheet molding compound in the design of its new cars. However, Opel executives expect renewed use of sheet molding compound when a new polyurethane recycling program is introduced.

Opel's DFE program has other significant elements. First, Opel designers will use recyclable material wherever possible. The material of choice is polypropylene, which costs less, is highly durable and recyclable, and has noise-reduction properties. Excluding fibers, polypropylene accounts for 68% of the plastics in the Astra model. Second, Opel will attempt to phase out polyvinyl chloride (PVC). This effort is driven by the refusal of municipal incinerators in Austria to accept vinyl waste. Third, Opel is committed to DFD principles. A new GM clamp uses a wedge catch and tooth strip for the rear bumper mount, enabling easy access and disassembly in 20 seconds. As much as possible, assemblies are made of like materials, without metals, to be fed directly into the recycling process; a classic DFR consideration. For example, a rack-and-pinion assembly previously made from steel and rubber is now made from polypropylene elastomer. Nylon and PBT elastomer are used in a new gear box design. Finally, as mentioned earlier, recycled materials are used whenever possible. Old polypropylene battery casings are now widely used by Opel in wheel arch fender liners; compounds are upgraded with bumper production scrap and virgin polypropylene. Polyurethane seat foam is used in a soundproofing mat in dashboards; scrap for these efforts is supplied by industrial plants, dealers, and scrap yards.

Opel's experience points to the fact that a "green" vision is essential for DFE to succeed. Once long-range plans have been developed, all capabilities and research efforts must be brought to bear in an effort to implement those plans. Opel used classic DFE guidelines with regard to material selection, DFMA, DFD, and DFR.

4.5.3 DFE in the Auto Industry

Other examples of DFE include the efforts of Mercedes-Benz to implement a total vehicle recycling program with two main elements: vehicle design and vehicle recycling [9]. Highlights of the Benz design effort include choosing environmentally compatible and recyclable materials for components, reducing the volume and variety of plastics used, marking plastic parts with logos, and avoiding composite materials as much as possible. Plastic parts are used only when practical advantages are offered by their use, and DFD and parts sorting upon breakdown are also given full consideration. Mercedes-Benz started taking scrap cars back in 1991. Once dismantling and shredding has been completed, the material is placed in a smelting reactor in a process Benz calls "metallurgical recycling." Obviously, such an extensive DFE program must be extremely well thought out, planned, and implemented from the very beginning. Such is the challenge for the DFE engineer.

BMW of North America recently announced a pilot program to test the feasibility of nationwide recycling of BMW automobiles [18]; because of tough German environmental laws, the company already recycles cars in Europe. Targeting three U.S. cities, BMW will give owners a $500 credit toward the purchase of a new or used BMW for turning in a car to a dismantling center. Using DFD principles and more recyclable components in the original design, BMW hopes to increase the percentage of recycled car weight from the present 75% to 90% in the future.

The U.S. auto industry is also looking at ways to break down polymeric materials into their original chemical state through hydrolysis, methanolysis, and pyrolysis. During hydrolysis, superheated steam is used to depolymerize resins, resulting in reduction to monomer. The monomer can be recycled when the plastic is repolymerized. During methanolysis, methanol is substituted to depolymerize resins. During pyrolysis, plastics and rubber are broken down to fuel-grade oil and gas by heating in an oxygen-deficient environment, resulting in usable products [19].

For the present and future, U.S. automakers will focus on the following areas to recover automotive plastics: batteries, where 150 million pounds of polypropylene is now annually recycled from 95 to 98% of all batteries for reuse in new batteries and other products; RIM (reaction injection molding) auto parts, which are currently landfilled; radiator end caps, where an infrastructure to collect the caps, recycle, and perhaps reuse them in other applications is being established; and SMC (sheet metal composite) scrap, where new product alternatives from this scrap are being explored.

In Europe, bumpers from scrapped cars are already being recycled. Volkswagen blends reclaimed polypropylene with virgin materials and uses this in bumpers for new cars. BMW reclaims PBT/polycarbonate bumper material from scrapped cars and uses the recovered material in new auto parts. Through a pilot recycling plant, in partnership with a cement manufacturer and a scrap metal processor, Peugeot S.A. hopes to produce a granulate, cement-making compound from 7000 scrapped cars over 2 years [19].

4.5.4   Germany's Green TV and DFE

As stated earlier, Germany leads the field as far as environmental awareness and action, and, more importantly, environmental legislation geared toward the recycling/reuse of consumer products. German lawmakers introduced legislation requiring equipment makers to take back used TVs and other electronics items for recycling. And even though "the government cannot tell industry how to make its equipment," DFE/DFR is a logical option for German manufacturers facing a real problem.

To this end, the German Ministry for Research and Technology and five European set makers have entered into a project to produce a fully recyclable TV; a prototype was introduced in 1994. The companies involved are Grundig AG, Loewe, RFT (all German), Nokia (Finland), and Thomson-Brandt (France). The motivation behind this novel partnership stems from the large number of TV sets (4 million) that Germans discard annually. Naturally, the key areas under study are disassembly, recycling, and reuse of components. Another feature of this alliance is the division into different areas of DFE research. For example, Loewe is examining DFD aspects, Nokia the recycling of picture tubes, Thomson the construction angle, RFT the recycling of coils and other assemblies, and Grundig the selection of appropriate materials and production processes.

To harken back to the earlier guidelines, it is clear that one path to the solution of far-reaching and complex environmental problems is the idea of pooling resources from different areas of expertise. Teamwork will reap rich dividends, and the beauty of it all is that everyone benefits. The international flavor highlights an important point:

To coin a phrase: United we stand, divided we fall.

4.5.5   Grundig AG's "Environment Initiative"

DFE requires firm commitment and perseverance. Take the overall manufacturing philosophy of Grundig AG, a $2.8 billion consumer electronics giant, which recently launched an "Environment Initiative" program. The goals of this program run the gamut from selecting suitable recycled materials to environmentally sound production and packaging. Upper management stands firmly behind these goals; a specialized team examines total environmental compliance in the production of all items and reports directly to the company's leaders.

Grundig consciously undertook a program to eradicate environmentally unsound chemicals from its production processes. Solvent-free, water-based cleaning materials are used almost exclusively, and organic solvents and CFCs are no longer in use. All subassemblies and materials containing fire retardants are free of polymerized diphenyl ethers to prevent emission of cancer-causing dioxins and furanes. In addition, Grundig uses materials free of formaldehyde, asbestos, cadmium, and mercury. At Grundig, lacquering is performed by highly efficient robots and machines using water lacquers.

Finally, Grundig recycles electronic subassemblies such as TV tuners, remote control units, etc., in its own module repair center, which is the largest of its kind in Europe. After repair, the parts are ready for reuse.

The Grundig experience shows that an overall "green" vision must pervade all aspects of a company's operations for DFE to thrive. Unhindered by its large size, Grundig sets a DFE example for all the world to follow.

4.5.6   DFE in the Consumer Appliance Industry

Appliance makers are probably on the cutting edge of DFE practice today; the U-Kettle is such a good example of DFE design. Further, since the consumer appliance market is so aggressive, manufacturers must keep pace with new materials and find that happy medium between cost and performance to stay competitive. The same environmental pressures exist for appliance producers, and since plastics have a wide use in this industry, material selection is impacted not only by cost/performance factors but DFD, DFR, and DFE trends as well. In addition, CFC elimination and power consumption reduction dictated by current and forthcoming laws provide motivation to research and develop product designs and materials that are environmentally sound.

Refrigerator and freezer manufacturers face the problem of finding substitutes for CFCs, whose use in the United States was to be eliminated as early as possible. Previously, CFCs were used as efficient refrigerators and insulators, but then it was discovered how their release could destroy the earth's protective ozone layer. The challenge, then, is to come up with an environmentally friendly substitute that has the same properties. The search continues.

For refrigeration, hydrofluorocarbon 134a has no effect on the ozone layer and could replace the soon-out-of-favor CFC 12, but it exhibits incompatibilities with some refrigerator parts. By the same token, CFC 11, used as insulation in refrigerator walls, has no real "green" substitutes that possess the same efficiency and noncorrosive properties. Designers have a choice: For better insulation, make refrigerator walls thicker. Unfortunately, given the limited kitchen space of most homes, larger appliances are not likely to be very popular with the consumer. But the search continues.

Also ongoing in the refrigerator arena is the use of engineering thermoplastics rather than CFCs as insulators. Designed by GE Plastics, one DFE design uses molded-in logos for material identification, a minimum of different materials, and compatible materials for combined recycling. DFMA features include parts consolidation and snap-fits. Once again, we witness the successful marriage of DFD, DFMA, DFE, and DFR in a single product.

As a final example of this marriage, consider Cleanworks, a compact modular dispensing system for industrial cleaning fluids. Designed for Scott/Sani-Fresh International, Inc., by Bally Design, Inc., it combines DFD, material reduction, and reclamation and reuse of bottles and product disposal.

The system consists of six recyclable, blow-molded, polyethylene bottles. Each is capped with an injection-molded, polystyrene Venturi pump assembly, which delivers a precisely mixed cleaning fluid from concentrate. The pump aspirates the concentrate and mixes it with water flowing through the dispenser. Once the module is empty, Scott comes to pick it up and replaces it. The pump assembly is removed from the bottle, ground up, and recycled. The existing bottle is refilled and then capped with a new pump assembly.

During design, environmental concerns were at the forefront. Both the designer and manufacturer were concerned with the consequences of discarded refill bottles. The manufacturer (Scott) took it upon itself to retrieve the bottles and reuse them whenever possible. When that wasn't possible, Scott made sure a recycling mechanism was in place. As seen in other examples, the recycling factor drove the choice of materials in the original design. Also, Scott had to set up the collection and recycling infrastructure for the DFE objectives of the design to be satisfied.

Besides DFE concerns, engineers are sometimes faced with several other conflicting criteria. Take the case of microwave ovens. With power ratings for these units on the rise, manufacturers must turn to new, high-performance engineering plastics to be compatible in the units. Sharp Manufacturing Company of America boosted power in its Carousel 11 models from 800 to 900 watts. The impact on the design was that the control panel and door opener button had to meet a higher UL (Underwriters Laboratory) relative temperature index. Moreover, the panel had to survive a drop impact test after withstanding a temperature of 204°F in a chamber. Sharp researched several materials before settling on a flame-retardant PC/ABS blend from Miles, Inc. At 0.125 inch thickness, this material met all UL specifications. The designers also found high impact resistance, dimensional stability, and resistance to ultraviolet and fluorescent light in this material. Aesthetic considerations were also satisfied as the material provided excellent colorability through color match with other components. DFMA principles were put to good use in this design; the control panel had clips molded to its back side, which are put through slots, slid down, and locked into place. The panel is fastened to the oven with just one screw. Sharp's design philosophy illustrates the necessity to conduct painstaking research on different materials, but it also shows how challenging the design task really is: Conflicting as well as diverse constraints must be dealt with simultaneously. Fortunately, the designers have tools such as DFMA, DFE, and DFR at their disposal.

4.5.7   DFE in Packaging Design

Our final DFE application will highlight the efforts of the packaging industry [15]. As in other manufacturing and design operations, the design of packaging is driven by cost/performance. The added environmental constraint means that, for the most part, packaging must be reusable or recyclable and still maintain function. Surveys have shown that most consumers are very concerned about a package's disposability and these concerns influence their buying decisions. Along with recyclability, buyers want convenience, freshness, and tamper evidence in their packaging; designers of packaging materials are sometimes hard-pressed to strike a balance between those requirements.

Most of the design emphasis for environmentally sound packaging is currently in the rigid packaging area. The most obvious solution appears to be that classic DFE premise: Use less material. But as discussed earlier, source reduction alone will not solve all our problems. As designers battle with the answers, they seek answers to some of the following questions: Should some resins and additives be avoided because, more and more, municipal waste is being incinerated?; Should packaging consist only of single layers (as opposed to multilayers) to facilitate recycling?; Will there be enough postconsumer waste collection and separation to justify designs using multilayers (for nonfood packaging) with recycle layers? The designers hope a public consensus on the matter of solid waste disposal will provide answers and help them shape their designs accordingly.

Current green practices in package design include the use of fewer materials, recyclable materials where possible, elimination of heavy-metal dyes, pigments, and inks, and the elimination of multimaterial lamination from packaging. Package designers are moving toward lightweight designs; structural layers get thinner as materials improve. More precise manufacturing process control reduces the need for "overengineered" barrier and adhesive layers. Designers continue the development of thinner, lighter, and stronger containers; as an incentive, Du Pont now hands out packaging awards for environmentally sound packaging.

Proctor & Gamble (P&G) has explored source reduction in Germany, where it markets Lenor fabric softener as a concentrate. Consumers get a refillable bottle into which they empty a small pouch of concentrate and enough water to produce 4 L of solution. P&G claims this effort represents a 90% source reduction when compared to throwaway detergent bottles. But interestingly, P&G credits a large part of the program's success to the green mentality of the German consumer. Says Tom Rattray, associate director, corporate packaging development, "They [German consumers] are personally affected by Germany's solid waste problem—most homes only get two smallish cans for trash per week and are acutely aware of their landfill limitations—in a country that incinerates 3.5 times more of its solid waste than we do."

In the recycling arena, P&G once again has made significant inroads. It uses postconsumer scrap in the manufacture of its detergent bottles; P&G uses in excess of 100 million pounds of high-density polyethylene (HDPE) a year for its bottles. On a limited basis, it bottles Spic'N Span, a detergent, in bottles made from 100% PET beverage bottle scrap. Other efforts include coextruded bottles from 20–40% reclaimed HDPE milk bottles. In the future, P&G hopes to double the recycle content of HDPE detergent bottles, even though the "recycle" content of a detergent bottle will never be 100% because of its contents. Detergents require at least a skin layer of the special grades of stress-crack-resistant HDPE that now make up 100% of most detergent bottles. So the use of recycled milk bottles in detergent bottles will, for the near future, require multilayer bottles.

4.6   SUMMARY

This chapter started with a gloomy forecast for the future of our planet. True, the statistics about environmental damage and neglect can be mind-numbing, but a look at the ingenious efforts of the designers and manufacturers using DFE indicates a sense of greater optimism. Hope does spring eternal from the human breast; in the case of DFE today, enough constructive work is being accomplished in the arena of environmental protection that there is indeed the feeling that human creativity will endure once more.

But this is no time to rest, for much work must yet be done. The fact remains: The "green" designs are here to stay, but so are the DFE challenges. Engineers must try to come up with a "green" angle on every project, although it is difficult to make much progress. They could design the perfect product for disassembly, but newsprint, aluminum, and plastic PET are the only recycling channels currently open in the United States. Channels aren't available to recycle most of the materials—the engineering plastics like ABS and polyurethane, that lend themselves to more precision molding—that designers work with. For designers, this is the biggest problem.

Recognition and definition of the problem derive half the solution; with slow and steady steps, we must capture the remainder. We have the tools at our disposal: DFE, DFR, DFMA, DFD, CE. With careful manufacturing and materials research, top management commitment, increased consumer awareness and support, production and design alliances, improved collection and recycling infrastructures, and above all dedication, patience, and creativity, the "industrial ecosystem" will be well within our reach.

4.7   BIBLIOGRAPHY

1.   Frosch, R.A. and Gallopoulos, N.E. Strategies for Manufacturing, *Scientific American*, 261, pp. 144–152, September 1989.

2.   Weissman, S.H. and Sekutowski, J.C. Environmentally Conscious Manufacturing: A Technology for the Nineties, *AT&T Technical Journal*, 70, pp. 23–30, November/December 1991.

3.   Wilder, R.V. Designing for Disassembly; Durable Goods Makers Build in Recyclability, *Modern Plastics*, 67, pp. 16–17, November 1990.

4.   Boothroyd, G., and Dewhurst, P., *Product Design for Assembly Handbook*, Boothroyd-Dewhurst, Warfield, RI, 1987.

5.   Arai, E. I., and Iwata, K. CAD System with Product Assembly/Disassembly Planning Function, *Robotics and Computer-Integrated Manufacturing*, 10(1–2), pp. 41–48, 1993.

6.   Ashley, S. Designing for the Environment, *Mechanical Engineering*, pp. 52–55, March 1993.

7.   Brooke, L. Think DFD!, *Automotive Industries*, 171, 71+, September 1991.

8.   Gosch, J. Leading-Edge Designers Design for Recycling, *Machine Design*, 63, pp. 12–14, 24 January 1991.

9.   Dvorak, P. Putting the Brakes on Throwaway Designs, *Machine Design*, 65, pp. 46–48+, 12 February 1993.

10.   Billatos, S.B., Guidelines for Productivity and Manufacturability Strategy, *ASME Journal of Manufacturing Review*, 1(3), pp. 164–167, 1988.

11.   Dambrot, S.M. Japan's Rising Interest in Ecotechnology, *Electronics*, 65, p. 28, 13 July 1992.

12.   Grogan, P.L. The Not-So-Original Disassembly Industry Biocycle, 33, p. 86, July 1992.

13.   Culp, E. Push to Recycle Auto Parts May Narrow Resin Use in Cars, *Modern Plastics*, 69(10), p. 68, October 1992.

14.   Gosch, J. Germany's "Green TV" Signals Trend in Set Design, *Electronics*, 65, p. 29, 13 July 1992.

15.   Schlack, M. Package Design Will Be Shaped by Solid Waste, *Plastics World*, 47, pp. 20–25+, September 1989.

16.   Lodge, C. Cost, Environmental Pressure Heat Up Plastics Designs, *Plastics World*, 50(9), p. 26, August 1992.

17.   Smock, D. Plastics in New Opels will be "Easily" Recyclable, *Plastics World*, 50(5), p.12, April 1992.

18.   Miller, B. SPI Launches Car-Recycling Program, *Plastics World*, 50(5), p. 16, April 1992.

19.   Kennedy, M. Plastics Group Drives Auto Industry Toward Recycling, *Machine Design*, 64(9), p. 14, 7 May 1992.

20.   Brooke, L. Take It Apart (Himont's Design for Disassembly), *Automotive Industries*, 171, p. 57–58+, June 1991.

21.   Forcucci, F., and Tompkins, D. "Automotive Interiors—Design for Recyclability," SAE Technical Paper 910852, 1991.

22.   Schofield, J.A. Programmable System Controls PC's Power Consumption, *Design News*, 49(5), pp. 149–151, March 1994.

23.   Murray, C. Snap-in PC Parts Aid Recycling, *Design News*, 49(9), pp. 45–51, May 1993.

24.   Machlis, S. A Peek at the Peripheral Grab Bag, *Design News*, 49(5), pp. 50–52, March 1993.

4.8   PROBLEMS / CASE STUDIES

4.1   Should design for the environment be done in parallel with manufacturing, testing, and service planning right from the beginning or once the product design is pretty well firmed up? Is the design time equivalent to product development time? Explain your answer with an example.

4.2   Evaluate at least three products and discuss new features that could be added to them to improve their environmental functionality. Current examples: computers that power down to conserve energy (e.g., Energy Star launched by EPA, IBM new power-down computers, Intelligent Energy Saver by

Unipower Corporation, Motorola 6805 8-bit microcontrollers, etc. [22]), computers and printers that can be designed for disassembly and recycling (e.g., snap-fits developed by Lexmark International [23] and Ecosys high-speed network printers made by Kyocera [24]), refrigerators that are CFC free, washing machines that use less water, clothes drying by microwaves, alkaline batteries that are rechargeable, and building heating systems that detect the presence of people and heat or cool as required.

4.3   PC boards are currently manufactured with solder alloys that contain lead. This is toxic and if not collected and recycled will likely end up in a landfill—and probably the groundwater. A number of alternatives exist, but none are a replacement for lead-based solder in all applications. One possible alloy is tin zinc (Sn–9% Zn). The problems with this alloy are: its melting point is 15°C higher than lead solder, and it probably requires an oxygen-free environment to work effectively.

a.   Examine the advantages and disadvantages of soldering processes and describe their environmental and health effects.
b.   Determine what design conditions are practical for soldering.
c.   Devise a means for providing an oxygen free environment for soldering PC boards. The best solutions will make it possible to rework existing equipment, that is, wave soldering, or surface mount conveyor / oven machines. Or, consider other alloys.

4.4   Polypropylene (PP) has recently found wide use in automotive applications because of new developments in polyolefin-based materials. These new developments have led to the creation of a family of PP products with a vast range of physical properties; from a DFE perspective, the most desirable of these properties is easy recyclability. Moreover, the DFE guidelines, stated earlier in this chapter, stipulate the use of fewer components, most of which can be recycled together. The "family" aspect allows use of different PP compounds in different applications for the same product (say, an automobile instrument panel), while the inherent compatibility of these compounds can also be exploited as part of a DFD/DFE strategy that produces large subassemblies that can be broken down with little handling. Composite structures may be obtained within the same part by using materials with very different properties. For example, glass-reinforced grades offer smooth processing, low warping, and extremely high rigidity, while filled grades possess high impact strength. If PP meets the performance standards for "fit and function" as dictated by a particular application, the designer's task has been simplified somewhat because with PP comes an in-place recycling infrastructure.

a.   Could other materials be used? What are the processing requirements and the limitations, if any, for these materials to be competitive with PP?

b.   The automotive industry is pioneering the application of DFE/DFD principles. For example, Himont Corporation [20, 21] uses the aforementioned PP properties to good advantage in DFD applications. Their PP instrument panel contains three primary elements covering a blow-molded duct system; the small number of large parts is made possible by the compatibilities discussed earlier.   Taken a step further, this means that during the disassembly operation, workers have only to remove three large parts as opposed to a number of small parts. Think of other parts of a vehicle (e.g., panels and consoles) and examine the impact of applying DFE guidelines to these parts if Himont's approach is followed.

4.5   Over the past several decades, industry has focused increasingly on developing technology that will reduce the environmental impact of consumer products and the production process. An industry that was not considered as a source of pollution is the housing industry. The truth of the matter is that the processing of building materials and the burning of fossil fuels for energy consumption put a strain on the environment. To help this industry, examine a residential house or an office building and find the areas of potential hazard to the environment. Consider, for example, the building's energy sources. Household energy accounts for about 20% of all U.S. emissions of carbon dioxide. It accounts for 26% of sulfur dioxide emissions and 15% of nitrogen oxides. More efficient use of energy could reduce these numbers through decreased consumption of fossil fuels.

a.   Is there a way to improve the energy efficiency?
b.   Are there alternative renewable energy sources?
c.   Could the building employ high-efficiency furnaces for part of its heating requirements and rely on the sun for the remainder?

FIVE

RECYCLING ISSUES

This chapter examines the methods that industry is currently using to help conserve our environment. Individual efforts in the recycling structure are also considered. Terminology describing these techniques is discussed in order to provide the reader with a clearer understanding of the topics.

5.1   INTRODUCTION

The rise in population of the United States and the increase of urbanization and industrial growth have resulted in an ever-increasing volume of waste that must be regularly collected, transported, and ultimately disposed of [1]. The average American discards total solid materials exceeding 600 times the person's weight during a lifetime. The disposal and accumulation of these solid materials has become a matter of increasing concern to American cities as well as the nation in general.

The national character regarding the solid waste problem was first recognized with the passage of the Solid Waste Disposal Act (PL 89-272) in 1965 [2]. This act authorized the Department of Health, Education, and Welfare to initiate and accelerate a national research and development program for new and improved methods of proper and economic solid waste disposal, and provide technical and financial assistance to state and local governments and interstate agencies in the planning, development, and conduct of solid waste disposal programs.

All residuals of the production process have only two possible ends. One is discharge to the environment, while the other is reuse, reclamation, or recycling. The discharge to the environment may lead toward endangering our own state of living, and therefore, more and more ways are being researched in order to increase the recycling availability. There are many definitions that describe recycling, but the following are the most common:

1. Reuse of products in the same capacity for which they were originally manufactured (bottles, pallets, crates, antiques, used houses, secondhand clothes).
2. The processing of residuals to produce the same raw material used in the initial manufacture of the final product (paper, glass, metals).
3. The alteration of the basic material of the residual to a completely different kind of material (using bacterial action to change cellulosic fibers in paper residuals to protein).
4. The release and use of energy contained in the residual (generating steam for the production of electricity by incinerating solid residuals).

By the end of the 1960s and the beginning of the 1970s, new recycling technologies had been developed and offered commercially. The social benefit of recycling is the value of the recycled material plus lower overall disposal costs minus the cost of collection and processing [3]. For example, substantial benefits and low cost can be obtained from recycled steel and aluminum products [4]. On the other hand, recycling smaller items with lower value, such as nails or newspapers, increases the net cost since the material has to be identified, transported, and separated.

Today's manufacturers cite that a desire to be good corporate citizens is one reason for the trend in design for recycling (DFR). Another point is environmental awareness on the part of the customer. The stick, however, is the possible federal legislation that could arrive much sooner if manufacturers ignore environmental concerns. The so-called "green" or environmental laws in Germany, the strictest on the European continent, hint at the kind of recycling rules that could be in the offering [5–8]. For example, the German government requires that auto manufacturers take back vehicles they manufactured at the end of the car's life and reuse all possible materials. In the United States, computer makers and three domestic auto manufacturers all are devising plans and guidelines for reusing the products coming off their production lines in order to prepare for compliance with the expected environmental laws in the future.

Metals and plastics are materials that are used most thoroughly in consumer products; they are virtually omnipresent. These materials make up more than 50% of our current MSW. Fortunately, they are among the most recyclable materials [9]. The major concern of DFR, currently still in the developmental stage, is therefore to decrease or eliminate the generation of such wastes, by reusing and refurbishing parts made of these materials and recycling them economically at their end-of-life.

5.2 COMPUTING CHANGE

It is estimated that at least 75% of product development and manufacturing costs is determined in the initial design stages. Design for recycling is intended to be implemented up front in the product conceptualization and design stages, according to the Congressional Office of Technology Assessment (OTA). Accordingly, the concept of DFR is becoming a formal discipline. A related idea, design for disassembly, also is becoming formalized. The two ideas complement each other nicely because the current state of the recycling art requires that products be easy to disassemble. Combined with other emerging areas of DFM and DFMA, the whole "design-for" realm is now increasingly referred to as DFX. For each DFX realm, design guidelines are set for the purposes of increasing the cost and time efficiency of X. These guidelines were generally devised during the 1980s to cope with keen international competition.

In general, DFR is a design process in which products' environmentally preferable attributes—recyclability, disassembly, maintainability, refurbishability, and reusability—are treated as design objectives [10–13]. As these environmental objectives are sought, it is important that the product performance, useful life, and functionality are maintained, and even improved. DFR makes good business sense because it lowers the costs associated with hazardous waste disposal while reducing the expenses associated with regulation compliance.

Until recently, product designers rarely thought about how their products would be managed as waste after their useful life was over, and the waste management providers tended to accept the composition as waste streams as a given. Therefore, it should be clear that, currently, DFR is still a developing systems method that promises to bring significant changes to future design proactive. In the United States, even the manufacturers (their names and methods will be discussed) that are most advanced in this field have just scratched the surface of the DFR effort. For instance, when a machine is returned to a manufacturer, workers remove and sort plastics by their types, they sort metals, and they collect wiring harnesses for recycling [14, 15]. This is the most common practice in the current industry. However, rather than this current cradle-to-grave practice, the Xerox corporation has begun to have its workers examine the xerographic module for possible refurbishing and reusing; this may occur several times before its end-of-life.

The idea, in general, is to first use parts having minimal wear and then recycle them when they cannot be reused. For plastic parts, recycling also means grinding and shredding components into particles or flakes, in some cases mixing them with new material, then manufacturing a new part. Though the process sounds simple enough, dismantling and identifying the materials in a worn machine may present special challenges. For example, just the adhesive on paper labels can contaminate the plastic to which they are bonded, which will make the plastic unfit for recycling. Further complicating matters are the thousands of different plastic available. Any one product may use dozens of different grades. How can a dismantler identify them all?

Molded-in labeling is a feasible solution to this problem. By using replaceable plugs, the generic material name and other information, such as percentage of reinforcing fiber, to assist recyclers could be molded in on every part. An example of this information one might find molded into an enclosure would be PPO + PS 20% GF, which denotes the makeup of polypropylene plus polystyrene and 20% glass fiber. In addition to the material identifier, developers might also mold in the manufacturer trade name and grade, such as GE Cycolac 2500.

In general, recycling is a dirty, unromantic business. For example, companies that operate municipal recycling facilities and automobile shredders try to cover their costs; they do not see themselves as environmental idealists. Instead, they try to figure out how to make a profit from consumer waste, which is their raw material. Materials (such as mixed plastic and newspaper) that have low market prices and are difficult to separate are not economically viable for recycling, and municipal authorities have to pay the recycling firms to take these items. Skeptics therefore believe that no one will ever make a profit from recycling, and that is why most automobile parts were not recycled before. Conventional wisdom holds that environmental considerations only add cost.

With the implementation of DFR, parts are designed to be easily disassembled and readily separated and sorted, which considerably decreases labor time to collect them. Although few companies are currently making a profit, they could expect a reasonable payback, both environmentally and economically, in the near future. The bottom line is that a positive payoff for any recycling effort is based on solid engineering planning and design. There is no doubt that applying design for disassembly and recycling rules in the design phase of a product pays off with a high end-of-life value.

5.3   HAZARDOUS MATERIAL CONTROL AND RECOVERY

Hazardous material control and recovery costs have always been extremely high. These economic factors, as well as the strong public concern for the human health and environment, have added up in favor of hazardous waste reduction.

5.3.1   Waste Reduction

Many hazardous waste problems can be avoided at early stages with the proper utilization of waste reduction [16]. Waste reduction refers to source reduction, which relates to less waste-producing materials in, and less waste out. It also includes treatment processes like incineration, which can reduce the amounts of wastes requiring ultimate disposal.

5.3.2   Hierarchy of Waste Reduction

There is a hierarchy of waste reduction, ranging from simple, to those that involve relatively drastic measures [2,17,18]:

- Increased diligence in housekeeping, such as using minimal amounts of water for washing equipment.
- Substitution of less hazardous material.
- Recycling and reuse.
- Process modification.
- Disposal.

Waste reduction can be achieved by careful inventory management, modification of manufacturing processes, volume reduction, and on-site reuse and recycling. The most effective waste reduction, however, can be accomplished by considering it at a very early stage of development or redesign, and at every single step of manufacturing. Figure 5.1 illustrates a waste reduction process in chemical manufacturing.

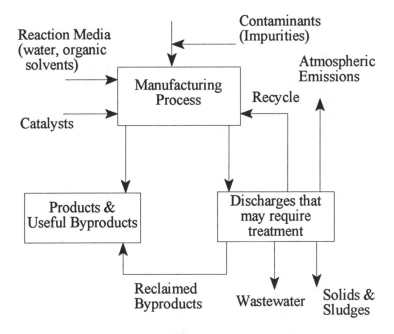

Figure 5.1 Chemical manufacturing process considering discharges and waste minimization.

5.3.3   Recycling

Several categories of substances are commonly recycled, among them metals and compounds of metals. Recycled inorganic substances include alkaline compounds, acids, and salts. Most of the organic substances consist of solvents and oils like hydraulic and lubricating oils. Catalysts are recycled in the petroleum industry.

Three major steps are used in converting waste oil to a feedstock hydrocarbon liquid for lubricant formulation. In the first phase, water and contaminant fuel are removed. In the second phase there is vacuum distillation processing, and finally, in the third phase, the lubricant oil stocks are separated from a fuel fraction and heavy residue.

Metals may be recovered by electrochemical reduction, in which a direct current is applied between electrodes immersed in the waste water.

Phenol, an industrial chemical with many uses, is recovered from wastewater by solvent extraction by the variety of water-immiscible organic solvents. Recycling of the waste water depends upon the characteristics of the same, and its intentional use. Sedimentation and filtration help remove solids. Trickling filters and activated sludge treatment reduce biological oxygen, and nitrogen can be removed by identification processes.

5.3.4   Waste Treatment

There are three major levels at which waste treatment may occur. The three levels are referred to as primary, secondary, and polishing. Primary treatment is capable of removing by-products and reduces hazard, but it is usually regarded as a stage of preparation for future treatment. Secondary treatment detoxifies, destroys, and removes hazardous constituents, while polishing is related to wastewater treatment. Figure 5.2 illustrates all the major phases of the waste treatment.

5.4   MATERIAL LABELING AND IDENTIFICATION

Material labeling and identification is a very important aspect that contributes to recycling. In the U. S. Environmental Protection Agency (EPA) hierarchy, reuse is given the highest priority, then recycling, and then reclamation or recovery. This standards also include an identification of materials by coding or marking with labels [19–21]. This helps to better explain how various parts can be taken off cars, sorted, and classified for recycling. In addition, alternatives to dumping shredder residue or fluff in the landfill can be evaluated and approved.

The German automaker BMW gives an excellent example regarding material labeling and identification. Of the production for the new 3 series, 81% of used materials are recycled. A special coding system is being used by the BMW designers in order to identify different components of the final product. Major plastic

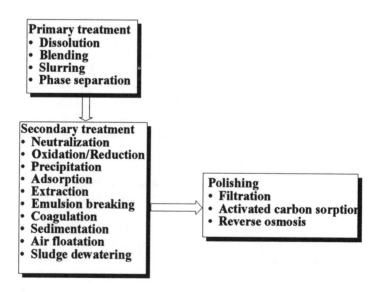

Figure 5.2   Major phases of waste treatment.

components in the 3 + 5 touring model are labeled by color-coding green for pure plastics, while blue is used to represent recycled plastics. The goal is to reduce the number of different plastics, avoid toxic and hazardous substances entirely, and mark all polymer parts that weigh more than 100 g. BMW also uses the color-coding system to identify fluids for easy separation and economical recycling.

Mercedes-Benz and Opel are other German companies that utilize material labeling and identification. Common procedure for these two companies is to use standard codes, such as the case of "PUR," used to label plastic components made of polyurethane.

5.4.1   Coding of Recyclable Materials

The recapture of recallable plastic from the waste stream will help lower the amount of waste going into landfills. The current plastic recycling technique of segregation by general families does not go far enough. The use of bar-coding plastic parts will supply important information about a plastic's composition. Bar codes and other scanning techniques can be standardized for the plastics industry.

In order to optimize the recycling process, plastics must first be identified and removed from the waste stream. One possible method is to incorporate a bar code on the plastic surface. This will allow the plastic to be sorted through the use of laser scanners and robotics. The bar code or Code 128 is derived from the Serial Item and Contribution Identifier Code. This formatted symbol is officially known as the SISAC symbol, from the Serials Industry Systems Advisory Committee. It is a computer-readable code containing alphanumeric characters. The bar code had its first significant application in 1973 in serials and publication tracking and data collection. The bar-code symbol, which is comprised of bars and spaces (normally with an aspect ratio of length to width of 1:5), is read through the difference in reflection of a scanning beam. Therefore, sufficient contrast is necessary between the bars and spaces.

The bar code can contain much information about the product to be separated. Examples include composition, age, environmental hazards, and additives. This information is valuable in determining the method to reclaim and recycle the material.

The challenge lies in this implementation of bar-coding plastics in consumer products to facilitate recycling. The procedure must be both efficient and economic so that the industry will embrace the concept or use it if mandated by the government.

Currently, municipal material recycling facilities sort waste using a high degree of manual labor. This is not an expedient approach to dealing with the amount of waste taken in on a daily basis. This method also makes the segregation of different plastics limited to a few categories. These categories contain material about which the facility knows very little in terms of its history or composition.

Municipalities generally find it more cost-effective to haul consumer waste in bulk to landfills. This holds true for most waste with the exception of appliances, motor oil, and batteries. These items are collected at request or dropped off by citizens at designated areas. The municipality will store them at a collection site until the quantity is sufficient that the cash from the metals/plastics will cover the cost of transportation and labor.

Scanning devices are available from several companies offering a wide variety of options at a wide price range. These can be combined with vision systems and robotics devices to perform the segregation of plastics into more distinct categories. Once broken down by chemical families, the plastic will be more valuable to industry as reground material.

5.5   METALS AND DESIGN FOR RECYCLABILITY

Metal recycling is not a recent development brought on by the current drive to recycle everything. Until about 1970 metal recycling being carried out was purely for profit. Many metals cannot be recycled for a high profit because of the many complications involved with the process.

The number of landfills in the United States has declined by 50% from 1979 to 1986, driving disposal costs as high as $150 per ton in many parts of the country. Clearly we must begin to reuse more and more of our wastes in order to preserve our dwindling supply of landfills. Metals, which make up a significant portion of the everyday waste stream, constitute 13.7 million tons each year, of which 1 million tons are recycled.

Every metal has its own properties and uses, for which it is assumed that recycling can be carried out in a rather simple manner. However, metals are difficult to recycle because of the chemical makeup or shape. For example, grindings are harder to recycle than large chunks of near pure metal. The dilemma is to identify the general problems associated with recycling metals as well as problems that apply to the recycling of a specific metal.

5.5.1   Applications in Metal Recycling

Recent surveys of recyclers yield a long list of problems encountered in the metal recycling business. Often kept secret or private, the problems encountered cover a wide range, from keeping equipment running to metal separation.

As with plastics, a major problem in metal recycling is the purity factor. Nonmetal impurities present new characteristics to a metal's composition. Separation techniques are slow and generally done by manual labor. Presorting is mainly used to categorize metal parts by color, density, and metal type. Presorted items are usually compared to sample items displayed on a board or some other type of display device. Comparison of like parts mainly constitutes how metal parts are classified and sorted.

In today's recycling industry, the most common form of separation is shredding. A shredder is a large, heavy-duty machine that can swallow an entire automobile and hammer it into fist size pieces. Pieces can be magnetically separated into piles of ferrous/nonferrous material. The nonferrous residue is about 25% of the weight that generally enters a separation shredder. A combination residue is generally sent into a metal miner. The metal miner is a machine that can remove nonferrous metals based on specific gravity.

5.5.2   Steel Can Recycling in the United States

Today the American public uses approximately 100 million steel cans per day, all of which are recyclable. The handling of postconsumer cans is a major topic that is just recently being addressed. Steel cans are those food and beverage containers, and other household containers, that mostly consist of steel. Every time a ton of steel cans is recycled, 2500 pounds of iron ore, 100 pounds of coal, and 40 pounds of limestone remain in the ground. All steel food and beverage cans are made from the highest quality steel available; therefore they are an excellent source of scrap. About 500,000 tons of prompt container scrap returns to detinners from can companies and mills. Of this, about 3 million tons reaches consumers.

Seven criteria are involved in the steel can recycling cycle. They are:

1. Food, beverage, paint, aerosol, and general-purpose cans that all are 100% recallable.
2. Homes and institutions where steel cans are used.
3. Cans that have been used and are ready for recycling.
4. Commingled recycling bin for curbside collection.
5. Curbside collection truck transports steel cans and other materials for recycling.
6. Magnetic separation, employed to separate steel cans from other recyclable municipal solid waste.
7. Molten steel, ready to be formed to make other steel products.

The recycling of steel cans has been successful due to a continuing commitment from responsible management and through support from the Steel Can Recycling Institute, SCRI. The SCRI is a nonprofit industry-sponsored association with the mission of promoting and sustaining steel can recycling throughout the United States. It is mainly concerned with helping community leaders, recyclers, business people, and educators in developing steel can recycling programs and to promote them to the public.

The SCRI advocates the recycling of empty aerosol cans through traditional curbside and drop-off recycling programs. While more than 500 programs are currently recycling steel cans, the SCRI anticipates that this number will increase over the next few years. Recently, the following communities have begun or will begin recycling empty steel aerosol cans:

•   Akron, Ohio
•   Austin, Texas
•   Gainesville, Florida
•   Tacoma, Washington
•   Oswego County, New York
•   Springfield, Illinois

There are three ways for communities to institute steel can recycling efficiently. They include curbside collection, resource recovery plants, and drop-off buy-back centers [22–25]. Each of the collection systems is currently being tested throughout communities in the United States. Obviously the easiest way for individual households to participate in steel can recycling is through a curbside collection program. In a multimaterial curbside program, families simply separate recyclable material from their household trash. The SCRI projected that steel can recovery rose above 50% in 1994 [26].

5.5.3   Aluminum Can Recycling

The practice of recycling aluminum cans is perhaps the one the public embraces most. Part of this can be attributed to the 5-cent payback that the consumer receives in many states for each can that they return to be recycled. In general, aluminum cans are 20% cheaper to recycle than to make, and recycling requires only 5% of the energy [27].

A process that maximizes the amount of aluminum collected from recycled cans has been developed at Ohio State University. The current process entails first heating the cans to 500°C to burn the lacquer and to remove the paint coating. The shredded cans are then placed in a furnace containing a molten salt and are heated to 750°C. The aluminum settles to the bottom of the furnace, where it can be collected. Currently, about 20% of the aluminum becomes trapped in the salt mixture. The process developed at OSU involves passing an electric charge through the salt bath used in the melting of the cans. This enhances the aluminum recovery, allowing companies to recover most of the previously unrecoverable 20%. This new process not only increases the amount of recoverable metal, but it also increases the number of times the salt mixture can be used. This new process will enable companies to obtain even larger profits from this already strong industry [29].

5.6   PAPER RECYCLING

One of the easiest materials to recycle is paper. Paper represents approximately 37% of our nation's waste [27–29]. Industry's goal is to recover half of all paper used by the year 2000. Current projections show that at the current rate of growth, this is attainable [30]. The three largest categories of paper are:

•   Newspapers
•   Corrugated boxes
•   Office paper

Although paper recycling is helping to reduce the need to cut down trees, it will never completely eliminate the need. Paper is usually made from chipped wood, softened into a wet mush and formed into a thin sheet. Recycling repeats this process with the paper itself, removing the ink, glue, and coating. This process breaks down some of the fibers, and the addition of new pulp is required to maintain paper strength. So recycling paper will never eliminate tree cutting, but simply reduces it (paper recycling requires an input of a minimum amount of new tree pulp to ensure a complete product [27]).

5.6.1   Paper Recycling in the U.S. Postal Service

Perhaps the single largest generator of postconsumer paper is the U.S. Postal Service [31]. It generates 1.2 million tons of undeliverable mail and paper discards annually. Approximately 90% of this, or over 1 million tons, is either landfilled or incinerated. Efforts to expand recycling in the U.S. Postal Service have been initially established in the Northeast.   Areas such as Springfield, Massachusetts, and Hartford, Connecticut, are leading the way. Between 90 and 95% of all postal facilities in the Northeast are expected to have recycling programs in the near future. Although the program is being met with success in the Northeast it may not do as well nationwide. Many post offices generate a small volume of material and are geographically far from a major market. Therefore, a proposal focuses on a termed 80:20 rule, that is, 80% of the materials generated by 20% of the facilities. If the Postal Service establishes recycling nationwide, it will be a welcome source of feedstock for the paper recycling industry [5].

5.7   PLASTICS AND DESIGN FOR RECYCLABILITY

Plastics products include a huge variety of items that are useful, convenient, and inexpensive, but on the negative side is the fact that as their volume mounts, so does the volume of waste as they wear out. Many plastic products have such a brief life span that they are discarded almost immediately in the waste stream. In short, the recycling level of plastics products has not matched the growth of the industry.

For reasons discussed in the following section, the focus of recycling efforts in the United States has been on packaging products, which have been recycled in continuously increasing volumes, but still seem low in comparison with sales volumes [32,33]. Recycling of other plastics products is far behind, but improved technology is being developed all the time, and the plastics industry is committed to the continued improvement of the recycling program.

Perhaps what is needed as a basis for resolving the problem of plastics recycling is an awareness of the entire life span of a product, the design and marketing aspects as well as the disposal and recycling stages. Once aware of the recycling problems, a designer could enhance and facilitate recyclability by making certain adjustments at the design stage. Once aware of the quality of recycled materials, the designer would be inclined to incorporate them in new products.

5.7.1   Design Reference for Plastics Recycling

To promote plastics recycling awareness, an attempt is made here to develop a combined design reference using available literature on plastics and recycling and other statistical data from the American Plastics Council and R. W. Beck Associates [26,27]. The focus of the this effort is directed to seven major plastics families whose products show significant recycling activity and whose market is packaging.

The seven plastics families are all commodity thermoplastics [30]. They represent 97% of plastic packaging materials. They are polyethylene terephthalate

(PET), polystyrene (PS), high-density polyethylene (HDPE), low- and linear low-density polyethylene (LDPE/LLDPE), polypropylene (PP), and polyvinyl chloride (PVC). They are mostly used in producing bottles and containers and other products that have brief usage periods before they are discarded.

Recycling efforts have focused on these plastics for several reasons:

- They constitute a significant portion of the solid waste stream (10%).
- They are constantly entering the waste stream.
- They are a highly visible component of waste and litter.

Thus these plastics are the target of environmentalists and the public calling for their elimination; the proliferation of recyclable collection systems around the country; the relative ease of identification, sorting, and separation; and generally good market demand for the recycled materials. Statistics indicate that plastic packaging recycling has increased nearly threefold since 1989, and that the number of recycling firms has also increased substantially.

On the other hand, there is in effect no large-scale recycling of postconsumer engineering thermoplastics such as acetals, nylons, and polycarbonates. These products are in autos, refrigerators, stoves, and other applications. They have long lives and are not disposed of in the usual waste stream. Cost-effective sorting is a serious problem.

Postconsumer thermosets such as phenolics, epoxy resins, and polyurethanes are not recycled on a volume basis either. Thermoset products are disposed of through industrial or commercial facilities rather than the municipal solid waste stream. Recycling efforts involve manufacturing scraps and several experimental and pilot programs.

Since plastics properties and characteristics have been covered in detailed in Chapter 6, a combined design reference for plastics and recycled plastics products, which promotes an overall awareness of a product's total life span, is compiled in a matrix format and presented in Appendices 5A and 5B. The appendices show each plastic family, beginning on the design side with a resin's key characteristics, primary market, and product examples. It then proceeds to the disposal and recycling side, first indicating statistical information concerning latest sales recycling rates and then the quality of the recycled material, the possibility of its reuse in the original product, and end-product applications for the recycled material.

One conclusion to be drawn from this discussion is that plastics recycling in the United States is really only beginning. Much needs to be done, not only in the area of the packaging plastics, but also with those plastics whose products are not currently being recycled in any appreciable volume. Recycling problems would certainly be reduced, however, and the recycling level would rise if all phases of a product's life were coordinated, as promoted by the design reference shown in Appendices 5A and 5B.

5.8   ECONOMICS OF RECYCLING

The simplest measure of the acceptance of recycling is economical. No one is willing to recycle MSW if it results in a financial loss for their company. Strong demand for recycled products ultimately requires that these products, as in the cases of steel and aluminum, be cost competitive and high quality [34].

Through 1993, the prices for many recycled products remained low. However, in 1994, the recycling industry experienced a resurgence, as prices skyrocketed. Paper prices rose dramatically, aluminum improved significantly, HDPE benefited, and steel remained strong. Analysts see these improvements as a sign that recycling has taken a major step forward. Much of this success is due to political intervention. In October 1993, an order was signed mandating that the federal government purchase 20% postconsumer paper by 1995. Many states and municipalities followed suit, adopting the same guidelines. Also, an international agreement to cut primary aluminum product, particularly in Russia, fueled the aluminum market along with a rising demand [26].

Many factors contributed to the low prices for recycled goods in the early 1990s. A rapid expansion of recycling programs resulted in an enormous supply of goods that the market could not handle due to an underdeveloped capacity to utilize recycled goods. Also contributing to the low prices was a sluggish national economy. It is unlikely that these factors will hit simultaneously again, allowing current prices to remain firm in the short term (the trend in current market has the price of aluminum essentially the same value for some years [26]).

5.9   PRACTICAL APPLICATIONS

5.9.1   Pollution Prevention: The 3M Company

3P Program. The Minnesota Mining and Manufacturing Company, also known as 3M, probably has the longest running and most successful environmental program in the United States. Pollution Prevention Pays, or the 3P Program for short, was initiated in 1975, and since then there have been 2700 successful projects for the first 15 years, $500 million in company savings, and a 50% reduction in pollution per unit of production.

The main goal of the 3P Program is to prevent pollution at the source in both products and manufacturing processes, rather than remove it after it has been created. There are four methods in achieving this goal: product reformulation, process modification, equipment redesign, and resource recovery. Responsible for implementing 3P within the company is 3M's vice-president for environmental engineering and pollution control (EE&PC).

When an employee recognizes a specific waste problem or pollution, a cross-functional team is organized to analyze the problem and develop possible solutions. The team itself might include employees from different divisions such as engineering, research, marketing, and legal. A simple submittal form is filled out and a proposal is forwarded to the affected operating division with a decision on whether to commit funds, resources, and time. One of the reasons behind 3P's success is the company's belief that the people who do a particular job are the experts. The program had no changes in its structure for the first 13 years of operation, which is another sign of its effectiveness. The following is an example of an approved 3P project:

The 3M facility in Alabama recycled cooling water that previously had been collected for disposal with wastewater. By reusing the water, the capacity of the planned wastewater treatment facility was scaled down from 2100 gallons a minute, to 1000 gallons a minute. The recycling facility cost $480,000, but 3M saved $800,000 on the construction cost of the wastewater treatment plant.

However, a new improved program had to be invented in order to strengthen the 3P principles with the future demands, this time called Pollution Prevention Plus, or 3P+.

3P+ Program. The goal of the 3P+ Program is to cut all releases to the environment by 90% from 1987 levels by the year 2000. This program is more structured than the original 3P, with waste minimization teams formed in every 3M operating division to identify source reduction and recycling opportunities and to develop plans to address them. One important feature of 3P+ is to denote emission credits to local permitting agencies, rather than sell them for its own profit. In addition, while 3P projects had to meet the same return-on-investment criteria as any other project, 3P+ takes into consideration more intangible benefits, like the enthusiasm and the morale of its employees.

The company invested $170 million in air pollution control equipment to reduce hydrocarbon emission by 55,000 tons, or by 70% from 1987 levels by 1993. Additional control equipment was installed at all plants emitting more than 100 tons of hydrocarbons annually, despite the fact that they meet the air quality standards.

Today, 3M is going even further by proposing for the development of a "beyond 3P" program that will primarily focus on the environmentally friendly issues. Environmental considerations can help the company become a lower cost producer. 3M believes that if its products compete on quality, performance, and price for the sake of environmental improvements, they will be "greener," and that would lead 3M toward a more competitive edge in the future.

5.9.2   A Leadership Role: Du Pont's Environmental Policy

Based on its waste minimization, leadership in plastic recycling, community involvement in environmental issues, and wildlife habitat enhancement, Du Pont has long been recognized for its environmental leadership.

Plastics Initiatives. After achieving industrial recycling successes, Du Pont moved into postconsumer plastics in 1989 with the Plastic Recycling Alliance, a joint venture with Waste Management, Inc. (WMI). The Plastic Recycling Alliance plans to establish five recycling plants in North America dedicated to certain postconsumer plastics, including polyethylene terephthalate (PET), high-density polyethylene (HDPE), and multilayered constructions. The joint venture was responsible for establishing one of the country's largest and most comprehensive plastic waste management systems. This network was expected to have a capacity to recycle some 200 million pounds of postconsumer plastics annually in the next 5 to 10 years. The following are the principles of this idea:

- Internal waste minimization programs help recover and reuse 1 billion pounds of in-plant polymer wastes and polymer intermediates annually, without compromising finished product quality.
- Packaging awards are designed to recognize innovations that reduce the amount of plastic packaging in the municipal solid waste stream.
- Employee programs make contributions to employee-designated charities for each pound of PET and HDPE bottles collected.
- A recycling pesticide bottles study examines the feasibility of recycling HDPE pesticide containers as a safe alternative to land filling.
- A beach clean-up program is designed to collect the plastic waste from beaches and recycle it into park benches.
- Food service recycling replaces the washing of dining ware.
- Motor oil bottle recycling collects used polyethylene motor oil bottles and returns these oil bottles with recycled contents.
- Automobile plastic components are also recycled by Du Pont.

CFC Reduction. A good example to prove Du Pont's good will toward saving the environment is its decision to discontinue its $750 million a year business in producing chlorofluorocarbons (CFCs) by the year 2000. It has been proven that these substances contribute in depleting the ozone layer and endangering our environment. In 1986, it was agreed, in the Montreal Protocol initiated by Du Pont, to cut CFC production by 20% in 1992 and another 30% by 1997. By the end of 1990, the company had invested $240 million in developing CFC alternatives, and it received the U.S. Environmental Protection Agency's 1990 Stratospheric Protection Award.

Education and Partnerships. Along with other member companies of the Council for Solid Waste, Du Pont has developed broad municipal solid waste educational programs. These programs are designed to educate the teachers regarding alternatives to reducing the solid waste, its recycling, and its impact to the environment by providing tools and funding.

Du Pont has entered many partnerships related to plastics, its recycling, and environment protection. Some of these partnerships are presented next:

- Partnership with the State of Illinois to develop commercial markets for PET and HDPE containers collected through curbside and drop-off recycling programs.
- Partnership with American National Can, a major producer of glass, metal, and plastic containers, to encourage the recycling of multilayered plastic bottles.
- Partnership with the Plastic Recycling Alliance and Occidental Chemical Corporation to implement an automated sorting system to separate PVC bottles from the mixed postconsumer bottles.
- Partnership with the Council for Solid Waste Solutions and the Flexible Packaging Association on recycling plastic structures and other forms of flexible packaging.

5.9.3   Johnson & Johnson: Worldwide Environment

Johnson & Johnson (J&J) is the largest and most comprehensive manufacturer of health care, pharmaceutical, and medical products in the world.

Environmental Policy.   This policy was developed in order to protect the environmental and natural resources related to all J&J operations worldwide through responsible management exercising excellence in environmental control.

Environmental and Regulatory Affairs Program.   This program covers the technical aspects of the environmental initiatives taking place at J&J companies worldwide.   The concept is incorporated in the J&J Worldwide Environmental Practices Manual by which each facility is evaluated to ensure that it is in compliance with local, regional, and national environmental laws and regulations.   The policies in this program are:

- The J&J Worldwide Storage Tank Policy establishes standards for proper management of storage tanks.
- The Biological Waste Policy requires all biological waste to be rendered noninfectious and unrecognizable prior to disposal.
- The Carcinogen Policy requires J&J facilities that use such materials in any of its manufacturing or laboratory areas to find suitable substances.
- The Chlorofluorocarbon Policy requires removal of CFCs from all J&J products as well as all manufacturing processes in compliance with the Montreal Protocol.
- The Product/Process Development Policy requires that an environmental impact assessment be performed on all research and development projects.
- The Domestic Hazardous Waste Treatment, Storage, and Disposal Facilities Policy requires all domestic affiliates to use only those off-site treatment, storage, and disposal facilities on the J&J approval list.
- J&J Real Estate Acquisitions and Divestitures Policy requires each J&J company to arrange for prepurchase environmental audit on all real estate property where environmental liabilities are identified.

Community Environmental Responsibility Program.   The goal of this program is to ensure that each facility worldwide operates at the same high level of environmental preparedness.   The four major phases of the program are:

- Phase I relates to the training of facility managers in crises management and is directed toward awareness, preparedness, and communication.
- Phase II, Facility/Community Communication, has the objective of establishing environmental leadership in employee/community communications and technical process technology.
- Phase III, The Marketplace, strives to make J&J the producer of environmentally natural products through their life cycle.
- Phase IV, Public Affairs Activity, strives to take a visible role in shaping

environmental initiatives in the civic arena.

Environmental Commitment.   J&J has learned from its Environmental Responsibility Program that it takes time, effort, and corporate commitment to get results. Communication is the driving factor for the entire program. As a foundation for environmental leadership in the future, the J&J environmental policy must be understood by all employees.

5.9.4   Choices for Consumers: The Proctor & Gamble Approach

Proctor & Gamble is committed to providing products of superior quality and value that best fill the needs of the world's consumers. As a part of this, P&G continually strives to improve the environmental quality of its products, packaging and operations around the world.—P&G Environmental Policy

P&G has pioneered a number of environmental concepts in the United States including:

- Concentrated detergent powders that require less packaging.
- Combination products of such a detergent with bleach or with fabric softener—another form of source reduction in packaging.
- Recyclable plastic packaging, and packaging with significant recycled contents, including 25% recycled content in plastic film packaging for disposable diapers.
- Composing of food waste, soiled paper products, and yard waste.

Organization and Policymaking. The company's efforts to develop environmentally safe packaging began in 1988 and became a priority for the packaging department. The goal was to make the environment part of the priorities among designers, along with containment, safety, and costs. Public education was also another goal for the company, to educate the employees in the organization as much as to educate the consumers. Empowerment and environmental policy were blended together.

Plastic Packaging Initiatives. In 1988, P&G launched the first major consumer product in a completely recycled plastic container. Today, Spic'N Span Pine cleaner is packaged in a 100% recycled PET bottle. In 1989, the Tide, Downy, and Cheer containers had 20–30% recycled content, which marked the beginning of an era of large-scale plastic recycling.

Another company innovation is the first coffee jar made from plastic that can be recycled. In 1990, the company began marketing Folgers instant coffee in a clear PET jar, in addition to offering the steel coffee can. While nonrecyclable packaging can yield significant source reduction, P&G sees this alternative as only an interim solution. In the long term, the company expects all of its products and packaging to be compatible with recycling or composting.

P&G doesn't stop there. It is involved in many more environmental actions today. It contributes in building an infrastructure; it forms partnerships with consumers, works with retailers, legislators, and activists, influences suppliers, and cooperates with other industries.

5.9.5 A Venture in Environmentalism: The National Polystyrene Recycling Company

The National Polystyrene Recycling Company (NPRC) was established in 1989 to create a program of responsible citizenship and to improve industry's dedication to manufacturing environmentally friendly products.

The NPRC's Los Angeles plant handles 13 million pounds of polystyrene a year. The Los Angeles school system uses 2.5 million pounds annually, and the NPRC facility is collecting from every school in the district. NPRC also sold 3 million pounds of recycled polystyrene feedstock in 1990 for application alone. It is also working to develop videotape cassettes, toys, and other household items.

The NPRC defines one of its challenges as that of changing polystyrene recycling from a supply-driven business to a demand-driven activity, and it is pushing for durable product manufacturers to become more and more involved in using recycled materials.

5.9.6 DFR at Dell Computer Corporation

Most companies are developing similar engineering rules of thumb, and a few are tackling other problems as well. For example, glue and adhesives that contaminate are slowly giving way to innovative assembly and manufacturing methods. Engineers at Dell Computer Corporation have been specifying an adhesive to hold small lights or LEDs in a front panel. Recently, Dell engineers were able to design small details into a molded enclosure so that the LEDs were able to snap-fit into place, eliminating the glue. This successful effort makes the enclosure more recyclable and cuts assembly time. What makes such substitution possible is the machining of details into a tool-steel hard mold and then locating the runners in the mold so that the details fill properly.

Partial credit for successful filling goes to new high-flow materials just reaching the market. In Dell's case, the material is a polycarbonate ABS alloy from GE called Cycoloy. Not surprisingly, material suppliers are helping to write the recycling rules. Engineering designers should be conversant with the material available in the market, and update this knowledge constantly. When a design calls for a particular strength, engineers can choose from special grades that might not recycle easily, or use a more readily recyclable material and design for a thicker part. There are thousands of plastics to choose from. It is important to focus on just a few with the right range of properties, and those that recycle well, if the concept of DFR is to be implemented on a design.

5.9.7 How Digital Corporation Uses Its Recycled Plastic

Just as important as designing for materials recovery is designing for the use of recovered materials. Developing designs that facilitate the disassembly and separation of product components is not enough; for true DFR, companies must incorporate recycled materials and components into their products. While the primary barriers to recycling are economic, the limited availability of high-quality recovered materials can complicate efforts to incorporate them into designs.

This issue is highlighted by a project developed by Digital Equipment Corporation to recycle plastics housings and keyboards from used Digital computers. Concerned about the disposal of its products in ever shrinking landfills, Digital established a process to recycle plastic computer parts. The goal was to reuse the plastics in new computer housings, but technical difficulties concerning polymer contamination made the original goal unsuitable. After researching the issue, Digital found Nailite Corporation, which recycles and reuses the polymer in roofing applications. Digital engineers developed a proprietary automated separation process based on specific gravity.

5.10   FREE MARKET RECYCLING INFRASTRUCTURE DESIGN

5.10.1   Environmental and Economic Issues

In 1993, the average American consumer created 4½ pounds of solid waste each day, of which a majority still ends up in landfills. This totaled 180 million tons of solid waste. The main components of solid waste by volume are 34% for paper and cardboard, 20% for plastics, 12% for metals, 10% for yard waste, 7% for glass, and 2% for rubber. The decomposition rate of materials in landfills has demonstrated to be very low. The average rate for biodegradable items such as yard waste and food products has been calculated to be approximately 25–50% over a 10- to 15-year period. Excavation by William Rathje, an archaeologist for the University of Arizona, has revealed readable newsprint from 1952. Items commonly thought to be very degradable, such as carrots, corn, and hot dogs, have been found identifiable after over a decade underground. This slow rate of decomposition may be in part due to the design of modern landfills. Landfills are mainly constructed to prevent contamination of the air and groundwater. Both air and water are prevented from reaching the solid waste by multilayered liners in the landfill. The environmental and potential heath hazards from leakage and emissions from landfills are placing pressure on their use.

The environmental issues are not alone. The cost to communities for collection and disposal of common waste is becoming a major factor to the popularity of the landfill. The United States will spend over $100 billion this year on environmental protection; that is about 2% of the GNP. This figure, which is greater than any other country in the world, is only expected to grow. By the year 2000, it is expected to comprise 3% of the gross national product. The defense budget in comparison is projected to be 3.7% in the same year. Most Americans pay for the disposal of waste through their local property taxes or by a fixed fee to a private collector. Paying a fixed fee does not provide incentives to modify consumer behavior in waste disposal. This flat rate is beginning to be challenged in small towns. On the other hand, unit-based pricing works where households are charged per unit of trash, for example, an established size of can or bag. In general, the number of municipalities with curbside recycling programs rose to 7265 in 1994, serving 7 million people more than the previous year [3].

There is no easy solution to the problem of solid waste disposal. The only certainty is that landfills, due to environmental and economic pressures, are becoming less popular.

5.10.2 Design Features to Minimize the Cost of Compliance

Far from being expensive taking apart old machines and reformulating the plastics of use in new H.P. equipment will save money...the [plastic] structures combine so well with our current formulations that the recycled materials are actually better than fresh resins.—Hewlett Packard Engineering

The prevailing thought for a long time in industry has been that recycling is cost-prohibitive. This idea was challenged more and more in the 1990s. That's what Hewlett Packard discovered after completing the first phase of recycling old computer equipment. H.P. believes that 5% of its plastic material needs will be made up of recycled material in the near future.

Many companies are trying to beat impending legislation that will require that their products must be recyclable. European manufacturers are among the leaders in including recyclability into product development. For example, in the mid 1990s BMW planning includes opening an innovative R & D recycling facility at Wakersdorf, Germany. The center is to concentrate on developing the techniques to dismantle and recycle products. Another example of Germany's DFR desire is the "Green TV" project. The project is a consortium of five European television manufacturers backed by the German Ministry for Research and Technology. The goal of the consortium is to develop a design that will facilitate easy disassembly of parts made from materials that are readily recyclable. This will help alleviate the number of discarded TVs, which in Germany alone is close to 4 million annually.

However, DFR sometimes clashes with modern design concepts. In the concept of DFM, where the number of parts in a product is reduced to simplify the design and decrease the assembly time, the engineer will face a dilemma. The product in DFM must be constructed in a way to facilitate disassembly for recycling, for example, with snapping fasteners. Snap-together parts usually do not lend themselves to easy disassembly. The engineer has to make the choice to support DFM with less parts or recyclability. Another dilemma of DFR involves reducing the amount of packaging material. This is a simple form of DFR that reduces the content of environmentally unfriendly materials in products. This has already been achieved significantly in the soft drink industry. Engineers have reduced the amount of glass in bottles by 43%, aluminum in cans by 35%, and plastic in bottles by 21%.

5.10.3 Existing Government / Business Infrastructures

The world leader in legislation toward recycling is Germany. Germany is the European equivalent to California, which has the largest amount of waste and the strongest environmental lobby. The German legislation to date regulates management and recycling of packaging in three phases:

- 1991—All transport packaging is required to be accepted back by manufacturers and distributors including items such as crates, drums, pallets, and Styrofoam containers.
- Manufacturers, distributors, and retailers are required to accept all returned secondary packaging including items like bulk cardboard boxes and blister packs.
- Manufacturers, distributors, and retailers are required to accept sales packaging. This includes all items necessary to transport products to the point of consumption. This consists of cans, plastic containers, foils wrapping, Styrofoam, and cardboard packaging.

In 1993, Germany targeted to have 50% of all packaging material collected, and of that, 30–70% to be recycled and reused. Future targets for as early as 1995 increase the percentages between 80 and 90%. Incineration to produce energy does not meet the legislation, so materials must be recycled. Failure to meet the requirements will result in stiff penalties.

German businesses have decided to face the new recycling climate together and have banded over 400 companies together to form the Duales Deutschland GmbH (DSD). DSD translates to the dual system of Germany and includes a few U.S.-owned businesses. The DSD has devised a green dot program. The green dot symbol is prominently displayed on products of the participating manufacturers. The dot indicates that the product can be returned to DSD recycling centers. A fee is charged to manufacturers, depending on bulk, who display the green dot. The fee is used to finance local waste collection and recycling programs. This sets up a nonprofit organization to handle solid waste. Consumers separate their solid waste and place it in specific yellow containers that are placed at curbside for collection. Contractors collect the yellow containers and deliver them to over 200 sorting centers. At the centers the items are again separated by the manufacturers and returned for recycling. To date over 16,000 companies have signed on to the green dot program devised by the DSD. One-fifth of all household waste will be collected, a large task with a large setup cost of $10 billion U.S. dollars and an additional $1 billion for annual operating expenses that ultimately will be paid by the consumer.

Cost is just one of the critics' claims against the green dot program. Some critics claim that the green dot program will not be self-financed as planned. They feel that the cost of collection and distribution will be greater than the value returned from the recycled material. Another argument against the DSD program comes from Germany's wine producers. The vintners feel this will only hurt their established practice of simply returning wine bottles to the manufacturers for reuse. The recycled glass from the green dot program is only a more costly extra step; however, vintners who stay with the old practice will not be able to display the green dot. Only time will tell if DSD and Germany's recycling efforts will prove fruitful.

Germany may be the leader but not the only player. Many countries are well along the path to setting up their own recycling infrastructure. Italy presently recycles

50% of all glass and metal drink containers. Switzerland has also reduced its drink container waste flow by reducing one half the weight going to landfills. The French have recently implemented legislation requiring industry to take back packaging waste for incineration to energy or recycling. This cost will be absorbed by a levy charged on packaging.

The recent trend is for governments to strike deals with companies to move further toward recycling goals that legislation would require. These companies, which play more of an active role, are more involved in controlling the time table for recycling to better suit their own needs.

5.10.4 Industry Working Together

One example of where leadership in industry is rewarded is in the United States. In Los Angeles, California, a smog-market scheme has been devised to reduced air emissions in four counties that surround LA. The plan, approved by the South Coast Air Quality Management District, employs competition to reduce pollution. R-E-C-L-A-I-M (the Regional Clean Air Incentive Market) involves over 390 local companies. These 390 companies emit more than 4 tons of nitrogen oxides and sulfur dioxides annually. These two pollutants are the major contributors to the smog problem in urban LA. Emission limits in these counties will be reduced on average 7% a year. By the turn of the century emissions of nitrous oxides should be cut by 75% and sulfur dioxide by 60%. Companies that lead the effort by decreasing emission rates faster than legislation will receive air pollution credits. The credits allow the release of pollutants in the future. Companies are then allowed to trade these credits with other businesses. Sulfur dioxide has been traded on the Chicago Board of Trade since March 1994, with an initial 150,000 units sold.

Another attempt of industry working together is European Recovery and Recycling Association (ERRA) in Europe. The main endeavor of ERRA is in setting up demonstration projects around Europe to raise public awareness. The demonstrations have found much public support, with some projects achieving 70% public participation. Another unexpected outcome is in the diversity of the participants; for example, the consumer research is handled by the Coca Cola company and the project manager is from Schweppes. All this private and governmental action is slowly creating an industry of itself. In Europe alone an industry of $100,000 to $200,000 is being established.

This is also becoming an industry in the United States. For environmental protection, an estimated $115 billion is currently being spent annually. The EPA is budgeted to spend just under $500 million on research and development alone. This new industry also encourages the development of new products. The EPA, for example, has a green lights program to which many companies have already committed. This program encourages the installation of new lighting technologies that will reduce energy consumption. Currently 10% of Fortune 500 companies have adopted this program. In one EPA analysis, companies that would convert to new lighting would reduce U.S. energy consumption by 11%. This could also reduce the amount of nitrogen, sulfur, and carbon dioxides each year. This would be equal to removing 42 million cars from the streets. In this scenario, both business and the environment will benefit.

Many areas of recycling are being developed in fields large and small. Polystyrene, which constitutes less than 1% of municipal waste, is a good illustration of a small area being developed. This procedure is sponsored by the Dart Container Corporation and is called the Michigan County Recycling program. Participants in the program separate polystyrene foam from other forms of waste and bring it to any number of 24-hour drop sites. Dart collects the foam and delivers it to the Mason, Michigan, recycling plant. The recycled polystyrene is then used in products ranging from egg cartons to video cassettes.

Manufacturers like Dart and countries like Germany are just a few examples of governments and businesses setting up recycling systems. Infrastructures for recycling to some degree are currently pursued in almost every industrialized nation. In the majority of cases, government working with business is the preeminent format [35].

5.11   BUSINESS AND CONSUMER FEELINGS ON RECYCLING

Concern about the environment is growing stronger every day, and all forecasts predict a continuing trend for the future. During the 1980s and early 1990s, strict legislation was passed mandating environmental protection. Companies that have failed to comply have been penalized with large fines. In reaction, companies are rethinking their environmental stance.

> *Our customers want our products to be recyclable..They want to feel that in purchasing our products that they are being environmentally conscious.*—NCR Corporation

Manufacturers are aware of consumer sentiment. In western Europe this attitude is prevalent in that customers are demanding that the products they consume will not hurt the environment. 3M operates in over 55 countries and recently surveyed over 2500 customers. 3M found that in picking a supplier, 20% of their clients thought it was most important to choose a supplier that cares for the quality of the environment. This compares to 17% who felt that monetary value was most important. Although manufacturers know the favorable position consumers have toward recycling and the environment, there still exists a prevalent negative view among business executives.

In a 1993 environmental survey of middle managers, published by Nexia US Ltd., 95% of all respondents agreed that business has a responsibility to protect the environment, and also agreed that they are willing to do their fair share. However, the managers have severe problems with the manner in which the regulations are devised and operated. These are:

- Sixty percent think environmental regulations are unreasonable, and 64% think penalties for noncompliance are unreasonable.
- Seventy-five percent feel there is too much paperwork involved in compliance, reducing profitability.

- Fifty percent think timetables for compliance are unreasonable.
- Fifteen percent believe the government plays a cooperative role with the business community.

Although most companies agree on the growing environmental consumer sentiment, this is not their prime incentive for going green. According to the survey, respondent say waste reduction and energy conservation were done mainly for profit improvement, recycling and pollution reduction were done for corporate responsibility, and hazardous waste disposal was done for regulatory compliance.

This survey revelation of an underlying view among business may be a transitional pain felt from the new environmental vogue. This antigovernment feeling should subside as business takes more and more initiative to change. It will be interesting to note the survey results at the turn of the century.

5.12   SUMMARY

We live in a world that is constantly changing. Pollution and disposal of products has increased, thus endangering the human environment and weakening the ozone layer. Fortunately, recycling helps reduce the dangers the world is facing from ever-rising amounts of disposal by transforming them back into use. In striving to achieve environmental responsibility, society faces daunting tasks. New technologies and infrastructures are constantly needed due to the changing requirements. That is where recycling leadership comes in order to pave the way for the future. There are many excellent cases in the world's industry today that qualify this leadership among the companies (Johnson & Johnson, Du Pont) presented in this chapter.

Manufacturers must address environmental and green design issues when approaching recycling. But in order to achieve the best recycling results, manufacturers should not be the only ones to play the role. Consumers, the media, educators, the government, the environmentalists, retailers, and the industrial ecology also have to take a significant part. Today, recycling is a global issue in which we all have to contribute. It is the only way to preserve our environment, and that is why our goal should be to always work hard in order to improve it.

5.13   BIBLIOGRAPHY

1.   Steuteville, R. The State of Garbage in America, *BioCycle*, pp. 54–61, April 1995.
2.   "Technical-Economic Study of Solid Waste Disposal Needs and Practices," Clearinghouse, 1969.
3.   Biddle, D. Recycling for Profit: The New Green Business Frontier, *Harvard Business Review*, 71, November-December, pp.145-156, 1993.
4.   More Aluminum from Used Cans, *USA Today Magazine*, p.7, June 1995.
5.   Cairncross, F. How Europe's Companies Reposition to Recycle, *Harvard Business Review*, 70, pp. 34–45, March 1992.

6. Dennison, U., and Katrin, B. Germany's New Packaging Laws: The Green Dot Arrives, *Business America*, 113, pp. 36-37, February 24, 1992.
7. Germany's Package Deal, *Futurist*, 26, pp. 53-54, November 1992.
8. Gosh, J. Germany's Green TV Signals Trend in Set Design, *Electronics*, 65, July 13, 1992.
9. Staudinger, J.I. *Plastics and the Environment*, pp. 11–52, Hutchinson & Co., 1974.
10. Craig, N., and Savage, C.M. "Concurrent Engineering: Hype or Hard Work?" *ABB CE Nuclear Power*, Spring, 1993.
11. Dvorak, P. Putting the Brakes on Throwaway Design, *Machine Design*, 65, pp. 46-48, February 1993.
12. Grove, N. Recycling, *National Geographic*, 186, pp. 92-115, July 1994.
13. Lave, L., Hendrickson, V. and McMichael, F.C. Recycling Decisions and Green Design, *American Chemical Society*, 28(1), 1994.
14. Ehrig, R.J. (ed.) *Plastics Recycling*, New York, Oxford University Press, 1992.
15. Farrissey, W. Thermosets, *Plastics Recycling*, ed. R.J. Ehrig, Oxford University Press, 1992.
16. Manahan, S.E. *Hazardous Waste Chemistry, Toxicology, and Treatment*, Lewis Publishers, 1990.
17. Henstock, M.E. *The Recycling and Disposal of Solid Waste*, University of Nottingham, 1974.
18. Hepler, H. C & D Waste Recycling: Raising Consciousness, *American City & County*, 109, pp. 32-42, January 1994.
19. Baker, E. *Industry Shows Its Stripes: A New Role for Bar Coding*, American Management Association, 1985.
20. SISAC, Serial Item Identification: "Bar Code Symbol Implementation Guidelines," Technical Advisory Sub-Committee of SISAC, 2nd ed., 1992.
21. Sobczak, T. *Applying Industrial Bar Coding* SME, 1st ed., pp. 48–119, 1985.
22. Pattison, S. Repaving, Reworking, Recycling, *Consumer Research Magazine*, 18, September, pp. 60-64, 1993.
23. Quimby, T.H.E. *Recycling—The Alternative to Disposal*, Johns Hopkins University Press, Baltimore, 1975.
24. "Resource Recovery and Waste Reduction," U.S. Environmental Protection Agency, 1975.
25. "Resource Recovery," Midwest Research Institute, 1973.
26. Beck, R.W. 1993 National Postconsumer Plastics Recycling Rate Study. 1993.
27. American Plastics Council. *Plastics in Perspective*, 1994 ed.
28. Rosenberg, J. AF&PA Surveys U.S. Paper Capacity, *Editor and Publisher*, January 21, pp. 31–32, 1995.
29. Seonghoon, H. An Economic Analysis of Household Recycling of Solid Waste: The Case of Portland Oregon, *Journal of Environmental Economics & Management*, 25, pp. 136–146, September 1993.

30.   Burnett, R. Engineering Thermoplastics, *Plastics Recycling*, ed. R.J. Ehrig, Oxford University Press, 1992.
31.   Steuteville, R. Pay Dirt at the Post Office, *BioCycle*, pp. 77-79, June 1995.
32.   Steuteville, R. Year End Review of Recycling, *BioCycle*, pp. 30–32, December 1994.
33.   Stilwell, E.J, Canty, R.C., Kopf, P. and Montrone, A. Packaging for the Environment, Amacom, 1991.
34.   Ashley, S. Designing for the Environment, *Mechanical Engineering*, March 1993.
35.   Carbone, A. Industry and the Environment: Making Business Part of the Solution, *Magazine of the American Scene*, 120, pp. 32-34, March 1992.
36.   Nonthreaded Fasteners, *Machine Design*, 66, June 1994.
37.   Owen, J.V. Environmentally Conscious Manufacturing, *Manufacturing Engineering*, October 1993.
38.   Rhode Island Solid Waste Management Corp., "Materials Recycling Facility," Annual Report, pp. 1-6, 1993,
39.   Shriver, B., Beiter, B., and Ishii, K. Characterization of Recycled Injection Molded Plastics for Material Life Cycle Analysis, *International Journal of Environmental Conservation Design & Manufacturing*, 2(4), pp. 13-18, 1993.

5.14   PROBLEMS / CASE STUDIES

5.1   Design and analyze new uses for recycled materials from municipal recycling programs, such as glass and ferrous metals. In order to understand and determine the benefits and economic advantages of recycling glass and ferrous metals, the processes involved should be understood. Glass is composed primarily of silica sand, along with feldspar, limestone and natural soda ash. This mix is combined with cullet (recovered crushed glass) and placed in a furnace to be melted. The molten mixture goes through refining, conditioning, forming, and annealing processes to reach the final product. Ferrous metals such as steel and cast iron go through a shredding process to make the material more manageable (the shredding process reduces the size of the scraps by a system of cutting, tearing, and shearing). After shredding, separation (of foreign materials), cleaning (of any organic coatings or other combustibles), and melting processes are carried to reach the final product. Give detailed definition of necessary processes and equipment and consider the real-world problems of recycled materials, such as material contamination during collection process.

5.2   A large retailer often ships merchandise from store to store depending upon the area in which the merchandise sells best. The company utilizes corrugated cardboard containers to transport the merchandise between the stores. It is the policy of the company that once the merchandise has been unloaded at the receiving store, the boxes are to be "broken down" and stacked for recycling collection. Is this the best policy with regards to recycling efforts? If not, what suggestions would you make to change the policy?

5.3   A materials recovery facility (MRF) projects to take in 10,000 tons of paper products in the coming year for recycling. The facility estimates that 50% of the paper products is newspaper, 30% is cardboard, and the balance is mixed paper. If the current average processing costs (including labor costs, etc.) and market prices are as given in the table that follows, is it economically sensible for this MRF to remain in operation?

	Newspaper	Cardboard	Mixed Paper
Processing Cost	$30/ton	$45/ton	$35/ton
Market Price	$50/ton	$25/ton	$20/ton

5.4   The U.S. Environmental Protection Agency (EPA) estimates that the amount of municipal solid waste (MSW) that has been recovered via recycling has grown from 7% in 1960 to 22% in 1993. However, the amount of MSW discarded has nearly doubled, from 82 million tons in 1960 to 162 million tons in 1993. The EPA estimates that by the year 2000 the national recycling rate will reach 30%. Assuming the same annual increase in MSW generation as was observed for the last 3–5 years prior to 1993 (1.5%), what is the difference in the amount of materials discarded in 1993 versus the projected amount in 2000? Plot your results. What do these results suggest about recycling efforts?

5.5   In recent years, increased efforts have been devoted toward waste reduction due to our troubled environment. Possible solutions proposed have included reducing the production of environmentally hazardous materials, incineration, and recycling. Implementing these solutions has been extremely difficult because of the high costs and other economic constraints. To ease the implementation process, the free-market recycling (FMR) infrastructure discussed in Section 5.10 must be redeveloped for the twenty-first century. For this FMR infrastructure to work, four phases must be integrated: collection, sorting, reclamation, and end use.

a.   Develop an FMR infrastructure for plastics and show how the four phases can be integrated.
b.   Would design practice that minimize the cost of compliance change or stay the same (e.g., would systems that identify and separate materials be highly automated, unmanned, etc.)?
c.   What are the economic and engineering issues needed to promote the new FMR infrastructure?

5.6   Many businesses generate large volumes of waste. Highly effective recycling programs can be developed to collect these waste from similar businesses on a routine basis. Examine some of the local businesses in your area and describe the type of waste generated and the possible ways of collecting and sorting waste for possible recycling. Also provide some guidelines to help these businesses to be environmentally conscious and actively participate in recycling programs. Suggested examples include retail businesses, restaurants and bars, office buildings, tire retailers, wood and building materials users, concrete and asphalt, etc.

5.7   A portion of this chapter was devoted to design for recycling (DFR). Cited throughout the text were examples of DFR techniques currently being implemented by industry. Think of 10 common products and name the drawbacks they have for recycling. How might they be redesigned using DFR principles?

5.8   Waste reduction is not only an industrial concern but also a domestic one. As a group project, generate a short report of the possible things you can do to reduce the amount of waste that you generate.

APPENDIX 5A. DESIGN REFERENCE FOR PLASTICS AND RECYCLED PLASTICS PRODUCTS

Resin	Key Characteristics	Primary Product Market	Product Examples
HDPE	Tough, flexible, translucent; good moisture barrier; excellent chemical resistance; organoleptic; impact resistant	Packaging	Milk & water bottles; base cups; pigmented bottles; film
LDPE	Clarity; chemical inertness; sealability; ease of processing; strength; toughness; flexibility; decrease in melt index increases strength properties while decreasing ease of flow. Increase in crystallinity increases stiffness, chemical resistance, barrier properties, tensile strength, and decreases impact strength, tear strength, and stress crack resistance.	Packaging	Grocery and garbage bags
LLDPE	Improved tensile, puncture resistance, impact, and tear properties. Outstanding ESCR, low temperature impact, warp resistance. High crystalinity produces a stiffer product and higher melting point than LDPE	Packaging, construction	Food & produce wrapping, dry cleaning, Grocery and garbage bags, piping
PET	Clear, tough, excellent barrier to gases like oxygen and carbon dioxide. Unique ability to prevent escape of carbon dioxide makes it ideal for use in soft drink bottles. Quite versatile because it can be converted into amorphous or crystalline products, and because of excellent physical properties. Good mechanical, electrical and chemical properties. Crystallinity ranges from 0% to 60%, degree of which affects chemical resistance, fiber ability, thermal stability, and water sensitivity.	Packaging consumer products	Soft drinks, liqueur, food, toiletries, and health product bottles

APPENDIX 5A. Continued

Resin	Key Characteristics	Primary Product Market	Product Examples
PS	Amorphous, brittle until biaxially oriented, then becoming comparatively flexible and durable. Versatile, low production costs, easy processing, excellent thermal and electrical properties. Grades: (1) crystal is transparent, can be injection molded or extruded; (2) impact is opaque but much higher impact strength. These grades degrade in sunlight; (3) expandable PS.	Packaging, numerous consumer products	Egg cartons, meat & poultry trays, fast food packaging & utensils, cassette cases, appliance cabinets, VCR's, decorative panels)
PP	Moderate cost, versatile, can be tailored to many fabrications methods and applications. Excellent chemical resistance; low density and highest melting point of the volume thermoplastics; excellent electrical insulator; very good moisture barrier but poor oxygen barrier.	Packaging furniture, auto parts	Juice & syrup bottles, batter cases, film
PVC	Second largest volume thermoplastic in U.S. Basically unstable, but most versatile of all plastics due to use of plasticizers and additives. Innumerable applications. Unplasticized is flexible. Good resistance to hydrocarbons and most aqueous solutions. Moderate heat resistance, good chemical and flame resistance; excellent clarity, puncture resistance, and "cling".	Packaging, construction, auto parts, many more	Bottles, packaging foam sheet and pipes, siding, coatings, sealants, auto upholstery, auto body parts, hand bags, tapes, sports equipment, medical gloves, carpet backing, many more)

APPENDIX 5B. DESIGN REFERENCE FOR PLASTICS AND RECYCLED PLASTICS PRODUCTS (SALES AND RECYCLING STATISTICS ARE IN MILLIONS OF POUNDS PER YEAR)

Resin	Products Recycled	1993 amt - Recycled - Sales - Recycling Rate	Quality of Recycled Material	Possible Reuse in Original Product	End Products
HDPE	Bottles; base cups	- 450.2 - 4,243 - 10.6%	Varies with level of contamination; uncontaminated material is competitive with virgin resin	25% blend with virgin resin for new base cups and household chemical bottles	Base cups, household chemical bottles, duck nests, divider sheeting, wind barrier
LDPE	Grocery & garbage bags	- 88.3 - 4,593 - 1.9%	Same as above	Used in blend	New bags and films
LLDPE	Grocery & garbage bags	Consolidated with LDPE in Beck report (see above).	Same as above	Used in blend	New bags and films
PET	Soft drink & custom bottles	- 1,598 - 447.8 - 28%	Minimal contamination results in very pure recycled material	Purest grades used in 25% blend with virgin resin to produce new soft drink bottles.	Soft drink bottles, fiber fill, shoulder pads, geo-textiles, carpet backing, strapping, sheeting for packaging

APPENDIX 5B. Continued

Resin	Products Recycled	1993 amt - Recycled - Sales - Recycling Rate	Quality of Recycled Material	Possible Reuse in Original Product	End Products
PS	Single use disposable applications; meat & poultry wrap, fast food knives, forks, cups, salad boxes, hinged containers, clamshell packages	- 35.6 - N/A - 1.5%	Some change in properties but similar to standard commercial injection molding grade materials. Presence of pigments restricts end-use to applications where color or color variation is acceptable.	No, due to FDA regulations	Office items; desk trays, waste baskets, rulers, card files, coat hangers, flower pots, serving trays, packaging fill, void fill, thermal insulation
PP	Bottles Battery cases	- 13.6 - 1,639 - 0.8% - 191.3 - 52 - N/A	Excellent, but diminishes as level of contamination rises	Blend with virgin resin	New battery cases, geo-textiles, industrial fiber

APPENDIX 5B. Continued

Resin	Products Recycled	1993 amt - Recycled - Sales - Recycling Rate	Quality of Recycled Material	Possible Reuse in Original Product	End Products
PVC	Bottles, other packaging	- 5.5 - 717 - 0.8%	Quality must be determined by heat stability. If defective, plasticizer must be added to recycled material. Properties must be characterized. Finally, properties of recycled material have to be adjusted by adding modifiers, stabilizers, and plasticizers for tailoring to specific product.	No	Pipes, fencing, non-food bottles

PLASTICS FOR ENVIRONMENTALLY
CONSCIOUS ENGINEERING

In the 1960s and 1970s new plastics seemed to reach the market almost every week. They came with family names like polyethylene, polyester, acetal, and acrylic, and they earned public recognition under trademarks such as Plexiglas and Kevlar. These high-strength, lightweight materials were molded, extruded, and stamped into just about any shape or product conceivable.

Now with the twenty-first century on the horizon, scientists and engineers alike are challenged to discover ways to *unzip* these polymer chains so they can be made environmentally safe. This chapter provides some basic information required to better understand how these versatile materials can be environmentally designed for the future.

6.1   INTRODUCTION

There are many reasons why the world utilizes the benefits of plastics to meet our everyday needs. Many common items such as detergent bottles, milk bottles, and new technology products, for instance compact discs, are quite commonplace when it comes to plastics.

There are many advantages to using plastics over their metal counterparts. The main advantages include the fact that plastics are corrosion resistant, light weight and have diverse properties [1–4]. Some plastics are rigid while others are flexible; some are tough while others are soft. Another advantage of plastics is that they are cheaper than most metals. Their low cost and relatively unlimited availability may be the main reason why industry is so abundant with plastics.

Indeed, there are many types of plastics in industry. With new chemicals being added to existing plastics, such as fillers and other reinforcing materials, there is a great concern with how they can affect the environment once their useful life is over [5–7]. There are so many types of plastics one cannot begin to understand how many types are being produced today. Eventually, plastics will be tailor-made to fit a specific use. This will expand their already increasing number.

In the following discussion, it is very important to provide a proper description of what each element is and how it is utilized for commercial use. It is necessary to understand plastics concepts and applications in order to be able to understand why an environmental issue exists and why it is such a complex problem. Research is needed to explain the many applications of plastics that are in use today. With this in mind, the chapter is broken down into two parts. The first is a detailed description of the families of resins, polyethylene, polyvinyl chloride, polycarbonate, polypropylene, ABS, reinforcing materials and fillers, and plastics compatibility. The second focuses on the environmental issues that surround the use of such materials. Recyclability, biodegradable additives, and processing effects on the environment will be discussed, followed by an example of how automobile companies are facing this enormous challenge.

6.2   THE FAMILIES OF PLASTICS—RESINS

The term polymer refers to materials made up of very long chains of molecules that are composed of many smaller molecules or elements. Interaction of these long-molecule chains is responsible for the properties we associate with plastics, such as flexibility, extreme plasticity, and low permeability.

There are many types of plastics commonly used today. Properties of these plastics will be discussed as well as their applications for their use. Two general categories of plastics include:

1. Plastics that, when a molded sample is heated in the absence of oxygen, soften more and more until they are near liquids. These plastics are known as *thermoplastics*. As in the general definition stated in the beginning, thermoplastics are made up of long chains of molecules that are not chemically joined to each other [10].
2. Plastics that, when a molded sample is heated in the absence of oxygen, either crumble, melt, decompose, or otherwise do not fully become near liquids. These plastics are known as *thermosets*. Thermosets are chemically made up of long chains of molecules that are linked to a number of neighbors, creating a three-dimensional web.

Plastics or polymers are synthetic materials based on the chemistry of carbon [2,4]. Polymer molecules are made up of numerous identical "mer" units that are linked together into long chains. These chains can be separate or linked. Each mer unit is based on carbon atoms. On a larger scale, polymers can be identified as materials that are easily melted and formed into a finished product with the application of heat. Plastics in use today contain some of the largest molecules known to science, yet they are individually far too small to be seen even under the most powerful microscope. Because of this, and because there are so many types of plastics, industry classifies plastics according to their properties and color [9].

Resins are the most versatile of all chemical compounds, and their industrial applications continue to expand. Prior to the introduction of resins, the term resin was applied exclusively to particular "sticky" substances in the form of yellow-brown deposits that extrude from certain species of trees [8]. As the knowledge of resins began to increase along with the new compositions by which they could be made, more and more products were classified as resins.

By mixing these resins together, new products with totally different properties began to emerge. Base resins are produced from either natural or synthetic materials. They are often used as coatings, films, or threads. Some resins contain monomers that can be combined to produce polymers. Most resins are only soluble in hydrocarbon solvents [9].

Types of resins include general-purpose, rigid, dispersion, and solution. General-purpose resins are homopolymers that are mixed with plasticizers to produce a flexible material. Rigid resins are primarily homopolymers that produce a stiff material in the absence of plasticizers. They are used primarily in construction. Dispersion resins can be either homopolymers or copolymers. Particles of dispersion resins are suspended in fluids and used as coatings. When heated the resin bonds to the coated material. Solution resins are copolymers used to produce continuous films.

A strong argument exists that resins and plastics are the same. The difference is that resins are primarily used as substitutes for a natural product in coating compositions whereas plastics are those compositions that involve a molding operation. The following outlines the typical resins used in industry today.

6.2.1   Natural Resins

Spirit-Soluble Resins [9]

Balsams: Fragrance is derived from evergreen plants. Used in medicine, as a flavoring, or in varnish.

Turpentine: Clear, volatile, water-immiscible hydrocarbon fluid; exudate of living and the heartwood of dead coniferous trees. Used as solvent or for cleanup work.

Mastic: Adhesive, exuded by bark of evergreen shrub. Used in varnish, lithography, and dentistry.

Dragon's blood and gamboge: Pigment, evaporated sap from Garcinia tree. Used as organic coloring in watercolor paints.

Damar: Wear resistant; from West Indies tree. Used in varnish/lacquer for high wear resistance.

Sandarac: Brittle, slightly aromatic, some transparency; derived from Australian tree of pine family. Used in incense or varnish.

Lacs: Shellac; secreted by various scale insects. Used as shellacs.

Oil-Soluble Resins

Rosin: Hard, brittle resin. Remains after oil of turpentine has distilled from crude turpentine. Used on violin bows, gymnasts' hands, and in ink.

Copals: Fossilized resin, derived from tropical trees. Used in varnish and lacquer.

Amber: Fossilized resin, from tree that has lost its volatile components after millions of years of burial. Used for jewelry, decorative material.

Oriental lacquer: Lacquers originated in China; they are a fast drying, high gloss varnish. Used for protection and decoration.

Cashew-shell nut oil: Oil excreted from cashew nut. Used to make resins.

6.2.2   Synthetic Resins

Resin Adhesives

Plywood adhesives: Plywood is made of a large number of thin layers all held together with plywood adhesives.

Wood impregnation: Saturating wood with a resin can make it fire resistant.

Resins for Textile Applications

Continuous coating: This process deacidifies textiles. Sometimes used to restore paintings.

Impregnation of fabrics: Saturate fabric with resin to strengthen fabric.

Resins Possessing Ionic Charges

Ion-exchange resins: Chemically, ions are reversibly transferred between an insoluble solid (resin) and a fluid mixture. Used in water softening, and metal recovery.

Resins for treating paper: Used for bond paper due to elastic properties and adhesive ability.

Monomer resins: Copolymer of various monomers with acrylic acid.

Coating Resins

Phenol-formaldehyde resins: Heated mixture of phenol and formaldehyde. Used in pot handles, bottle caps, and switches.

Alkyd resins: Coating polymers; hard and brittle. Used in solvent-based paints.

Polyester resins: Long-chain polymers; hardness and resistance to burning. Used in boat hulls, old protective armor, cars.

Epoxy resins: Epoxy groups blended with other chemicals to form strong, hard, chemically resistant resins. Used as strong adhesive and as enamel coatings.

Urea–formaldehyde resins: Heated mixture of urea and formaldehyde; not for outdoor use; can be foamed. Used for colored countertops, cushion material.

Melamine–formaldehyde resins: Heated mixture of melamine and formaldehyde.

Used in differently colored countertops.

Polyethylene: Polymerization of ethylene. Used in making translucent, lightweight, and tough plastics, film containers, and insulation.

Coumarone-formaldehyde resins: Mix of coumarone, derived from coal tar, and formaldehyde. Used in paints and adhesives.

Silicones: Basic structure of oxygen and silicon atoms. Used for oils, polishes; resistant to heat and water.

Fluorine resins: Nonflammable resin. Used in nonstick pans such as Teflon.

Polyamides: Compound with two or more amide groups. Used in nylon.

Polyurethane: Produced by polymerization of hydroxyl OH and NCO group from two different compounds. Used in elastic fibers, cushions, insulation.

Polyethers: Contains ethers that will separate, and some will evaporate into the air. Used as solvents.

Acetyl resins: Contains the acetyl group, which can also be found in acetic acid.

6.3   POLYETHYLENE (PE)

One of the more common types of plastics in use today is known as polyethylene. Polyethylene, sometimes referred to as polythene, has been in existence since 1879, but had no commercial value then. It did not become popular until the 1930s when it was discovered that polyethylene contained excellent electrical insulating properties. Today, polyethylene is used to such a wide extent due to its chemical composition. The properties of polyethylene provide for it to be rather flexible. It contains the ability to withstand low temperatures while retaining flexibility. It has low water absorption and has an excellent resistance to chemicals. This may be the main reason why polyethylene is so popular in the food and beverage industry.

Polyethylene is commonly divided into density subgroups one through four, with one being the lowest density. Increasing the molecular weight gives an increase in impact, tensile, and tear strength, environmental stress crack resistance, fusion temperature, heat seal range, and lower brittleness temperature. These properties are the result of a tendency for polyethylene to become more crystalline as the density increases. This is from a reduced branching in the atom chains of higher density polyethylene. As density or crystallinity is reduced, increases can be expected in impact resistance, cold flow, tackiness (blocking), clarity, and tear strength. There will also be decreases in stiffness, tensile strength, hardness, abrasion resistance, and brittleness temperature [3].

Very-low-density polyethylene (LDPE) is used for shrink wrap, diaper films, and health care products. An example of the low-density form is a clear sandwich bag or the transparent squeeze bottle used for cleaning up liquids or stains. High-density polyethylene (HDPE) would obviously mean that the molecules are closer together, and hence the material is more rigid. Examples of this include a clear plastic milk bottle, chemical containers, truck bed liners, pallets, small boats, and large-diameter industrial piping.

Ultra-high-modulus polyethylene fibers exhibit the highest specific strength of any manufactured fibers. These fibers can be used in composites in the same manner as carbon fiber. This application of polyethylene has great potential for tennis rackets, bicycles, and high-performance automobiles. One of the current problems with the application of polyethylene fibers is creating acceptable adhesion with the surrounding matrix and a low service temperature. These two factors are slowing the application process at this time.

Blown film, cast film, blow molding, and injection molding are some of the techniques used to form products from polyethylene.

6.4   POLYVINYL CHLORIDE (PVC)

Another abundantly used plastic is polyvinyl chloride, or PVC. North America produced approximately 33% of the world's 18,633,000 million tons of PVC last year [11]. Polyvinyl chloride is unique because it consists of 56% chlorine by weight. Due to the large content of chlorine, PVC is often under attack. Despite public pressure, the PVC market has continued to grow by approximately 3% per year [12].

The individual responsible for discovering PVC is Henry Victor Regnault. He was a French chemist and physicist whose discovery of vinyl chloride came about in 1838. His other discoveries included dichloroethylene, trichloroethylene, and carbon tetrachloride. Regnault was born on July 21, 1810, and died, after a very influential life, on January 19, 1878. Regnault earned a place in the Inventors' Hall of Fame for all of his work in physics and chemistry [9].

The use of PVC went stagnant until World War II. This stagnation period was a result of the lack of technology in the material sciences. As the world of technology boomed for the material sciences, so did the synthetic polymers industry. The new technologies allowed synthetics to be produced inexpensively. The commercialization of synthetic plastics was also made possible due to the superior degree of property management compared to the plant- and animal-based polymers of the past [13]. During World War II it was soon discovered that PVC provided excellent electrical insulation for wiring. Today, it has become the most widely used material for electrical insulation.

PVC is a thermoplastic. The simplest thermoplastic structure is polyethylene. Polyvinyl chloride is made from an ethylene monomer [14]. The monomer is a chlorinated organic compound that can be polymerized into a useful synthetic. In this case, one chlorine atom replaces one hydrogen atom. This actually makes PVC a nonflammable polymer. Thus, PVC is made by starting with a polyethylene structure that is polymerized.

Polymerization is simply a reaction where small molecules react to form larger molecules. These large molecules are called polymers [9].

Vinyl chloride polymerization is accomplished by the following steps. The first stage is initiation. This process involves decomposing thermally unstable materials such as peroxides, percarbonates, and peresters. After initiation, vinyl chloride monomer units are added in a chain until the high-molecular-weight polymer is created [12].

The process of the monomer units joining in chains is exothermic. This usually occurs at 40–70°C. Note that "poly" means many. Thus, many vinyl chloride molecules make polyvinyl chloride.

The final resin particles average 100–150 micrometers (also called microns). After a suspension process, the remaining vinyl chloride monomer is removed under heat and vacuum conditions. The resin slurry is then centrifuged to remove the water [12].

Like its polyethylene counterpart, PVC properties include water, chemical resistance, and exceptional strength. PVC is a rigid plastic. If plasticizing materials are compounded with the PVC, it can become flexible. Certain chemicals such as modifiers can be added to PVC to make the material either soft and flexible or very tough and rigid. Since PVC resins are thermally sensitive, they require heat stabilizers to be added. These heat stabilizers are commonly known as plasticizers. Plasticizers allow flexibility and create a wide variety of PVC uses [12,15].

Polyvinyl chloride is more commonly used than any other plastic with the exception of polyethylene [12]. The major products of PVC include upholstery materials and waterproof fabrics such as shower curtains. Pipe fittings, chemical storage tanks, floor covering, packaging films, garden hoses, and plumbing materials are just a few applications of PVC.

6.5   POLYCARBONATE (PC)

Polycarbonate appeared in Germany around 1959. Its properties provide an excellent material for engineering applications. Polycarbonate has outstanding toughness, long life under extreme conditions, and excellent machining properties [16]. The characteristics of polycarbonate include excellent transparency, electrical properties, and superior dimensioning properties. Current applications for polycarbonate include cups, measuring beakers, baby bottles, measuring jugs, and thin film in electrical condensers. It provides an excellent substitute for light metals. Polycarbonate applications continue to grow, including moderate use in the architectural field, where it is now being put to use as bulletproof walls due to its toughness.

Some thermoplastics can withstand temperatures of 500°F and higher. They are stiff, strong materials that have outstanding chemical and wear resistance and good electrical properties. Many of the thermoplastic polymers are highly ignition resistant, having V-O flammability ratings at thickness down to 1/32 inch or lower without flame-retardant additives. Liquid crystal polymers, polyketone, polyphenylene sulfide, imides, and sulfone polymers, have comparable properties of high-temperature thermoplastics [17–19].

Several new technologies have been developed for the small but quickly growing rotational molding market. Rotational molding is a process in which materials such as polyethylene (PE), cross-linked PE, polycarbonate, powdered polyvinyl chloride, plastisols, nylon, and other materials are injected or poured into a mold cavity that is mounted on an arm that rotates the mold on two axes, thereby tumbling the mold, according to *FSP Machinery*. Growth in this market, estimated by some to be 10% annually, has been fueled by an increasing demand for parts such as large waste range of tanks. Shipments of major rotational molding resins were 404 million pounds last year, an increase of 26% from 1989, according to new data from the SPI's Committee on Resin Statistics.

U.S. Technology, based at Canton, Ohio, has developed an old series of plastic recycling niches, with an abundance of other environmental overtones besides recycling. Beginning with industrial scrap polyesters, polycarbonate (PC), acrylics, urea, and melamine, the company has the materials ground into blast media of different sizes and degrees of abrasiveness. Thermosets are ground into chips, for example, and PC is extruded into micropellets the texture of coarse sand. The blast media is used in dry paint stripping, and U.S. Technology then recycles the waste from this process into a line of cast polymer bathroom fixtures [12].

Commercial polycarbonate polymer blends were reprocessed up to five times to determine the influence that reprocessing has on their structure and physical properties. In injection molding the parts are usually wasted into scrap material because reprocessing is justified from an economic point of view, especially considering the high price of the engineering thermoplastics used nowadays. The commercial importance of polymer blends has clearly increased in recent years.

Economic recovery has rejuvenated the engineering plastics market, as healthy sales of autos, home appliances, and business machines has triggered demand for many resins. Consequently, engineering plastics prices are increasing, and lead times are beginning to extend. According to researchers at the Freedonia Group, North American demand for engineering plastics could rise by about 8% a year, reaching 2.6 billion pounds by the end of 1994 and almost 4.5 billion pounds by 2000. Of the engineering plastics, polycarbonate is expected to register the strongest growth, says Freedonia. Polycarbonate applications in auto and other transportation markets will drive demand for the plastic, which is popular largely because of its strength to weight performance.

Nikon's long-awaited F4 professional autofocus camera ($2472.50) is undoubtedly becoming the benchmark 35-mm SLR (single-lens reflex) of the 1990s. Nicely contoured, the F4 is wrapped in a stylish, functional polycarbonate shell that looks and feels great [20].

Better Quality Cassettes (BQC) is the first user of the new nonopening mold retrofit kits from Galic Maus Ventures for the injection molding of optical-grade polycarbonate compact disc (CD) substrates. BQC has improved its productivity so that it now has capacity for 12 million CDS per year [21].

Part of a special section on international air transport, the Airbus Industries A300–600s ordered by Thai Airways International are equipped with updated cockpit styling designed by Porsche. Polycarbonate panels cover the main instrument panel, and there is a padded frame around the glare shield [22].

At a recent news conference, GE plastics introduced its new high-flow grades of polycarbonate PC/ABS [23]. The firm's Cycoloy C2950 HF resin, a new high-flow, low-heat, flame-retardant PC/ABS, will be used for thin-walled computer and business equipment applications. Target applications of these new resins include microwaveable cookware, thick-walled connectors, extruded raceways, and headlamps [24].

China's plastics industry is expanding rapidly and is poised to enjoy growth in the future. Output of plastics products rose from 1 million tons in 1980 to 3.67 million tons in 1990, the highest in China's history. According to incomplete figures, there are over 220 plastic machine-building works in China, which produced about 180,000 units in 1991 [25].

Development of new polycarbonate compounds continues to expand the already broad base of applications. The manufacturers have announced new polycarbonate materials, including some with high flow characteristics, new blends and alloys, and six resins fabricated from recycled polycarbonate products. New alloys are being developed for customized applications that meet specific property requirements. Most of the new polycarbonate resins based on recycled materials that were introduced at the NPE contain at least 25% postconsumer content.

6.6   POLYPROPYLENE (PP)

Polypropylene, which is derived from petroleum, is made from carbon and hydrogen. Polypropylene is rooted in the discovery of propylene by Berthelot in 1869. The unpolymerized version of polypropylene was nothing more than a useless viscous oil. Polypropylene was first polymerized in 1954 by Professor Guilio Natta from an organometallic catalyst derived from titanium and aluminum. It was the first time a specific molecular structure was made to order [26].

Polypropylene is a highly versatile material. Polypropylene homopolymers and random and block copolymers can be produced for numerous and versatile applications. Its impact resistance and flexional strength make it suitable for chair shells, suitcases, television cabinets, toys, etc. Although it has a waxy feel like polyethylene, it cannot be scratched easily. It is the lightest of all thermoplastics with a specific gravity of 0.90–0.91. It exhibits good chemical resistance and is easily pigmented [27,28].

Polypropylene is extruded into pipes, films, and sheets. It is ideal for injection-molded items. Biaxially oriented, polypropylene film can be stretched both ways and is very tough. It is commonly used as cellophane and shrink wrap. Laminated onto book jackets, it can be bent thousands of times without cracking. Polypropylene has a higher melting point than polyethylene, so it can be sterilized and thus has many medical applications, such as tweezers, filters for artificial kidneys, and baby care products [28].

A unique property of polypropylene is that it can be molded into integral hinges that can be flexed millions of times on briefcases, suitcases, and toolboxes. Polypropylene fibers and filaments can be melt spun. They are stretched and annealed, made into straps for packing pallets, and woven into shiny sacks, ropes, pile carpets, and carpet backing.

Polypropylene is very common since it represents a major breakthrough for furniture design. Like many other plastics, molded polypropylene can be reinforced with glass fiber to increase its strength and heat resistance. With this treatment it forms the distributor belt guard in the Fiat 128 and the air grille on the Lotus Elite. Polypropylene can also be reinforced with asbestos. It is commonly found in automobile components including battery cases and instrumentation boards, and detergent compartments in dishwashers and washing machines. Additional uses include office equipment, packaging such as yogurt and margarine tubs, medicine bottles, and screw-on caps.

About 20% of polypropylene is made into a copolymer. Adding 2–5% ethylene increases clarity, which makes it easier to color. It also increases toughness and flexibility, but it lowers the melting point. Isotactic polypropylene is made by polymerization with organometallic stereospecific catalysts [28].

Polypropylene is commonly compounded with fillers and reinforcements such as calcium carbonate, talc, mica, and glass fibers. Blending it with elastomers makes it even more ideal for impact resistance and flex life. Other additives improve antistatic control, ultraviolet light resistance, long-term heat aging, radiation resistance, and flammability.

6.7   ABS, ACRYLONITRILE-BUTADIENE-STYRENE

ABS is a thermoplastic—a thermopolymer that is primarily made up of three petroleum-derived monomers, called acrylonitrile, butadiene, and styrene. It is a viscoelastic thermoplastic material whose properties are time, temperature, and strain rate dependent. When copolymerized with acrylonitrile alone, styrene produces a strong material with very good chemical resistance called styrene-acrylonitrile (or SAN). In the late 1940s, it was discovered that if styrene-acrylonitrile was polymerized with butadiene, in the form of fine particles, not only was the material's chemical resistance improved but the butadiene that was added increased its impact resistance. The polymer is composed of discrete rubber particles (based on butadiene) grafted with SAN copolymer dispersed in a continuous matrix of SAN. By varying the proportions of the three ingredients, as well as altering the molecular arrangement within the polymer, ABS can be produced with a variety of properties to suit the many different functions for which it was intended. ABS functions may include high impact, extra high impact, medium and low impact, heat resistance, and flame retardation.

The typical ratio of ABS consists of 20% acrylonitrile, 20% butadiene, and 60% styrene. Although ABS is a strong material, it will degrade under ultraviolet light, but this can be countered using a laminating acrylic film or by plating. Heat resistance can be added by adding alpha-methyl styrene as a fourth monomer.

ABS was the first plastic to imitate metals convincingly and can be chromium plated relatively easily. This process is difficult with most plastics. Such moldings provide lightweight replacements for electroplated metal parts on cars and cameras.

ABS is generally pleasant to touch, so that many high-gloss, smooth-finish products have been created from it. These items include housings for hair dryers, shavers, food mixers, and toys such as the famous Lego set. Refrigerator liners and door panels are overwhelmingly constructed of ABS in order to comply with new energy use requirements.

In the construction industry, ABS competes with high-density polyethylene, unplasticized PVC, and polypropylene in the replacement of cast iron piping and fittings. ABS can be processed in a variety of manners common to most thermoplastics. ABS can be injection molded, extruded, and machined and will readily reproduce etchings or engravings incorporated in the mold. Machining characteristics are similar to nonferrous metals, and many of the same tools can be used in most cases.

6.8 REINFORCING MATERIALS AND FILLERS

Growing industrial activities create a continual demand for improved materials, such as plastics, that will satisfy more stringent requirements, corresponding to higher tensile strength, modulus, thermal expansion, thermal conductivity, heat distortion, and temperature.

Reinforcements and fillers have always played an important role in the plastics industry. Polyvinyl chloride, polypropylene, and polyethylene have properties that will meet the requirements of high-volume end users, for which they have been sold and used as essentially pure resins. However, with price escalations and projected shortages in petroleum and resin feed stocks, there results an urgent need for widespread utilization of fillers that will reinforce plastic materials.

There are many types of plastic fillers and reinforcements in use today. The major reinforcements include high aspect ratio mica and other flake reinforcements, ribbon reinforcements, short-fiber reinforcements, and continuous-filament reinforcements.

Flakes, or platelets, represent a special class of reinforcing fillers for thermoplastics and thermosets. Mica is one such type of reinforcing filler that is relatively abundant and therefore the cost is quite low.

When reinforcing thermoplastics with high aspect ratio mica, great care must be taken to minimize flake breakdown during normal mixing, compounding, extrusion, or injection molding operations.

Considerable attention has been given to the use of glass flakes as reinforcing elements in resin systems. Glass flakes are mainly used for decorative purposes and for conferring special properties onto thermosetting resins, such as impermeability, scratch resistance, and chemical resistance.

Aluminum diboride flakes may be used as reinforcements in epoxy resin systems and phenolic resin [29]. Use of aluminum diboride flakes provides an unusually high degree of strength, stiffness, and modulus. Steel disks and platelets provide reinforcing, through an overlapping effect, where tensile strength will be dramatically increased. Alumina flakes provide plastics with high strength since alumina has a high tensile strength and modulus.

Another type of reinforcing filler is commonly known as ribbon reinforcements. These types of fillers, usually glass, will add to plastics properties such as mechanical properties (strength and stiffness), thermal properties (expansion and coefficient), chemical properties (durability and corrosion resistance), and transport properties (permeation coefficient).

Short-fiber reinforcements, such as asbestos, wollastonite, microfibers, and whiskers, will provide plastics with various forms of reinforcement characteristics.

Asbestos, which is an extremely hazardous material when airborne, is mainly utilized as a reinforcing filler due to its exceptional resistance to heat, oxidation, and ozone, and it is essentially insoluble in water. Although asbestos may be used with plastics, it makes it very difficult to feed directly into an extruder, injection molder, compression mold, or any other type of forming device. The main advantages to using asbestos as a reinforcing material is that it very significantly increases a plastic's modulus, heat deflection, temperature, and flexural strength.

Wollastonite, which is a nonhazardous mineral reinforcement/filler, provides plastics with exceptional tensile and flex strength, and exceptional moisture resistance. Under ultraviolet radiation and water emersion, the properties remain unchanged. The types of plastics that mainly utilize wollastonite include thermosetting polyesters, nylon, and polypropylene.

Microfibers, which are mainly composed of calcium silicate, are extremely useful in both thermoplastics and thermosetting plastic compositions. Microfibers will increase a polymer's tensile strength by two times and its modulus by five times.

Whiskers, which are microfibers that have been grown under controlled conditions, can be added to plastics to provide increased tensile strength, flexibility, and heat resistance.

Continuous-filament reinforcements such as fiberglass and high-modulus organic fibers are also fillers for reinforcing plastics.

Fiberglass, which is a very popular alternative to reinforcement, will increase tensile strength and heat resistance.

High-modulus organic fibers are used as a filler reinforcement (see Table 6.1) when lighter weight, more stiffness, and higher tensile strength composites are required. Kevlar is very popular in plastics reinforcement, along with Nomex, which

Table 6.1. Typical Reinforcing Fillers Used in Plastics

POLYMER	FILLER
Acetal	Glass fibers, TFE fibers
ABS	Glass fibers
Dially Phthalate	Glass fibers
Epoxy	Clay, silica, glass fibers, metal powders
Melamine formaldehyde	Flock, asbestos, fabric, glass fibers, cellulose fibers
Nylon	Glass fibers, asbestos
Phenol formaldehyde	Wood flour, asbestos, mice, glass fibers, cotton flock, metal powders
Polycarbonate	Glass fibers
Polyester	Clay, glass fibers, woven cloth, asbestos
Polyethylene	Asbestos, metal powders
Polyphenylene oxide	Glass fibers
Polypropylene	Asbestos, glass fibers
Polystyrene	Glass fibers
Polyvinyl chloride	Asbestos, clay
Silicone	Asbestos, glass fibers, quartz
Urea formaldehyde	Cellulose fibers, asbestos

provides flame-retarding attributes. Kevlar is mainly used for reinforcing elastomers, thermoplastic resins, and thermoset resins. High-modulus organic fibers provide plastics with a wide range of commercial and industrial applications. These applications include the tire and rubber market, aircraft, marine, sporting goods, and electrical industries.

6.9   COMPATIBILITY OF VARIOUS PLASTICS

Probably the most important requirement in mixing plastics together, along with any additives, is that it should be effective. This implies that it should be effective for the purpose for which it was designed, at an economic level. Improvements in one property can lead to deterioration in others and may have a negative impact on the overall performance of a particular design.

Complete compatibility, such as miscibility at the molecular level, and mobility or diffusibility of any additive molecules will occur when all the molecules of a system interact properly with each other. Miscibility simply means that the molecules are soluble and capable of mixing.

An estimated 33% of polymers are compounded in one way or another. To accomplish this task, there are two types of mixing. The first is distributive mixing. This type of mixing distributes compounds without the use of high shear stresses. Dispersive mixing, on the other hand, introduces high shearing stresses in order to break up cohesive agglomerated solids [28].

The compatibility and diffusibility of plastics in polymeric compositions is normally assessed by trial-and-error methods, which is a practice that is still likely to be used for some time in the future. The principal reason for this situation is mainly attributed to the lack of sound scientific methods. Useful methods are mainly used for solvent/polymer systems and may have limited applicability to the case of additive/polymer mixtures, where the polymer is normally the major element.

An example of successful plastics modification can be found in the polycarbonate industry. One of the most impressive characteristics of polycarbonate is its impact resistance, or toughness. The companies that use this resin decided to capitalize on the toughness properties. Their first question was, how can we improve the toughness properties without decreasing the high flow characteristics? The simple answer could have been to add more rubber. The problem, however, was that the rubber additive ruined the flow properties. After experimentation, however, the company discovered that the thermoplastic acrylonitrile-butadiene-styrene (ABS) could be the answer. This alloying process allowed flow properties to improve without any significant change in the toughness [30].

Returning to polymer compatibility, here are some examples. The case of polycarbonate can be used once again to gain valuable insight on this topic. Successful mixing has occurred using ABS to improve low-temperature ductility. PET has been a compatible mix that improves abrasion resistance. PBT can be used to improve the chemical resistance of a polymer. Thus, the compatibility of plastics is important to improving properties [22].

Most polymer mixing applications come from a specific project goal. A new taillight housing for Ford Motor Company, and computer business equipment housings that are UL-approved for flame-retardant capabilities are just two examples. These two specific needs led to successful polymer mixing experiments [31].

The idea of recycling has also led to new polymer mixing techniques. Recyclers are able to take bumpers from cars and recycle the material. This material is then supplemented with virgin material, stabilizers, and modifiers to obtain the desired physical properties.

Thus, in addition to compatibility, the number of additives available and the possible combinations in plastic systems are enormous.

6.10   GENERAL LISTING OF PLASTICS USES

A general listing of plastics and the major uses associated with them, both mentioned and not yet mentioned in this discussion, include:

Acrylics: Molding compounds, glazing.

Epoxies: Protective coatings, adhesives, laminating compounds, and composites.

Melamine-formaldehyde (aminos): Molding compounds.

Nylon: Filaments, fibers, bristles, engineering applications.

Phenol-formaldehyde (phenolics): Molding and laminating compounds, adhesives, cast sheets, rods, tubes.

Polyesters, saturated: Fabrics, food packaging, molding compounds, engineering applications.

Polyesters, unsaturated: Laminating and molding compounds, protective coatings, castings.

Polyethylene: Film, extruded tubing, sheets.

Polystyrene: Molding compounds, foamed resins.

Polyurethane: Protective coatings, adhesives, foamed resins.

Polyvinyl chloride: Film, sheets, extruded tubing, casting.

Silicones: Rubber compounds, molding compounds, protective coatings, paints, varnishes.

Urea-formaldehyde (aminos): Textile treating compounds, molding compounds, foamed resins.

Vinyls: Protective coatings, extruded tubing, film, molding compounds, foamed resins.

Appendix 6A provides a comprehensive reference to many of theses polymers and their properties including corrosive water and chemical resistance, tensile strength, impact strength, type, class and group, typical applications, limited oxygen index (LOI), ignition temperature, and heat of combustion.

6.11   PLASTICS, ENVIRONMENTALLY CONSCIOUS ENGINEERING

Following the discussion of the properties of the various types of plastics, a discussion of the recyclability of these materials is very important. Recently, much concern has been given to the use of plastics, because they do not biodegrade, resulting in plastics accumulating in landfills.

This has been the source of great concern, in recent years, that the use of plastics will have a negative impact on our environment. With pollution, ozone depletion, and global warming becoming so evident, today's industries have to begin to develop plans to recycle components that have outlived their usefulness. It is technically feasible to recycle, recover, and reuse all the discarded plastics that are used in packaging, such as polyethylene bottles. While there is a great concern over the disposal of plastic, plastics themselves do not directly effect the environment when deposited in a landfill.

The high stability of plastic essentially guarantees that there will be no interaction with the environment. However, plastics currently constitute about 18% (by volume) of all landfill material and this is likely to grow in the next 10 years as output grows and the amount of discarded plastics continually increases, resulting in larger landfills. It is the rapidly declining number of landfills that has caused great concern with the disposal of plastics. If all the plastics manufactured were used twice, meaning they were recycled once, the amount of discarded plastics would be reduced by 50%, but this may still not solve the plastics waste problem.

Industrial plastic waste usually consists of only one polymer type or grade. Because of high stability, properties do not decline during recycling and grinding. This is often the only operation required before the material is ready to be melted and processed into a finished product. Often, the recycled material is melted and blended with virgin materials, resulting in keeping 100% of the plastic's properties during processing conditions.

However, there are still several problems when it comes to recycling plastics [32,33]. Recycling of postconsumer plastics has begun only recently and was initially limited to rigid packaging materials such as polyethylene bottles (clear plastic bottles such as milk bottles, etc.) [34–37]. The problems for future recycling includes the primary areas of collection, sorting, and purification. The variety of different materials often used in one product also creates a problem. For example, polyethylene bottles may be clear or colored and may contain an additional five materials. High-density polyethylene bottles range from clear milk bottles to highly pigmented detergent bottles. The materials will usually end up at a recycling center where they will be wrapped up together. Automated separating/sorting facilities may be available to distinguish one type of plastic from another.

The value of clear, unpigmented resin is highest since it gives the user the widest range of application choices. For example, clear polyethylene can be mixed with other fillers and can be reused as a resin. Pigmented polyethylene resin can only be used as a filler for other products.

Still, other directions are being sought, for which the main concern is money. Many manufacturers have turned to using biodegradable polymers [38]. When these polymers were first introduced in the early 1980s, many actually did not biodegrade to provide a cleaner environment, as was advertised. Many of these resins left small pieces of conventional polymers after a biodegradable binding had dissolved. Coupled with this fact, these resins were not cheap. Costing as much as $10 per pound, they were also much more expensive than common resins that manufacturers were purchasing. While there were more manufacturers producing these biodegradable resins, it is still quite unclear whether companies would purchase them despite pressure to reduce the environmental toll caused by common resins.

The issue of biodegradability remains a popular idea despite the more expensive cost over conventional plastics. The physical properties of any plastic depend on the length of molecular chains that make up the material. The greater the length of these chains, the stronger and tougher the plastics. As soon as the chains are broken, the plastic begins to become fragile. If the process proceeds to a sufficient extent, it will become so brittle that it begins to break up under natural forces (forces such as wind, rain, etc.). As the plastic is reduced to smaller and smaller particles, it becomes more susceptible to biodegradation by microorganisms.

This chain breaking begins as soon as the plastic is exposed to solar radiation. Obviously, there is a time period before biodegradation occurs. This time period is dependent on the plastic's molecular weight. The point in time where a plastic's properties change is known as its critical molecular weight. Once the weight is reached, the properties of the plastic begin to make a remarkable change. The properties, such as time of biodegradation, can be determined in the manufacturing process.

The key to manufacturing a biodegradable plastic is the selection of the correct photosensitive group. This will provide a particular plastic with a desired rate of degradation. The ketone group is used extensively in this area. Ketones will absorb ultraviolet light. Using this property, the efficiency with which the plastic reacts with ultraviolet light can be altered. The biodegradation rate is proportional to the thickness of the plastic material. The thicker the plastic, the more ultraviolet light it can absorb. As a result, thicker plastics will have a longer life. The opposite will occur if the plastic material is thin. This process will be dependable for industrial plastics such as polyethylene and polypropylene.

The rapid growth that plastics have experienced, along with their expansion into new markets, has drawn attention away from people or groups that are concerned with environmental protection. Unfortunately, the term biodegradation has not been applied consistently, resulting in confusion among scientists and lay people alike. Deterioration or a loss in physical integrity of a material is often mistaken for biodegradation. Primary biodegradation is a biochemical transformation of compounds by microorganisms. Ultimate biodegradation of a material results in mineralization or incorporation into the microbial biomass. Mineralization of organic compounds will yield carbon dioxide and water under aerobic conditions or methane, carbon dioxide, and water under anaerobic conditions. The distinction between deterioration, primary, and ultimate biodegradation is important. Deterioration can result in a fragmentation of the plastic but will not necessarily remove the plastic from the environment. In fact, a reduction in particle size resulting from deterioration can result in a wider distribution of plastic particles in the environment [39].

Furthermore, attempts to engineer plastics with specific functional attributes, by increasing the durability and useful lifetime of a plastic, usually decrease the biodegradability of the plastics. Degradable plastics can be divided into several different categories based on the mechanism that is used to break down polymer chains. The Freedonia Group [40] listed a number of degradable plastics and estimated a percentage of their annual growth to be as follows:

- All plastics—4%
- All degradable plastics—16.8%
- Biodegradable—15.9%
- Photodegradable—14.5%
- Bio/photodegradable—18.2%
- Other degradable—54%

The wise use of the world's resources requires that we acquire knowledge of the effects of human activities on the resource base, as well as the overall quality of the environment. The ultimate cause of any resource depletion or environmental degradation is consumption.

6.12   ENVIRONMENTAL PROFILE

6.12.1   How "Green" Is Plastic Processing?

In terms of breaking down popular utilized plastics, examples are given as to how a particular product is produced and the toxic effects it will have upon the environment.

PVC Half-Gallon Bottle. Among all the plastics that are manufactured today (for common everyday purposes) PVC manufacturing requires the largest amount of process energy, followed by bottle manufacture and chlorine manufacture. In processing PVC into half-gallon bottles, the polymerization step accounts for the largest portion of air pollution, mainly in the form of hydrocarbons.
    Pure PVC is rarely used. Instead, it is formulated with heat and ultraviolet stabilizers, antioxidants, pigments, and plasticizers. PVC will soften at 70°C and begins to degrade at 150°C. This degradation produces hydrogen chloride. If the PVC should be completely degraded, half of its weight in hydrocarbon would be produced. Carbon dioxide and monoxide would also be produced in significant quantities [41].

High-Density Polyethylene Milk Bottle. The five manufacturing steps for one-gallon high-density polyethylene milk bottles leading to a production of 1000 containers will lead to both water and air pollution. Half of its air pollution is in the form of nitrogen oxides, while its waterborne wastes are primarily suspended solids.
    In wastewater treatment, there are two types of total solids. One type is the suspended solid, and the other is the dissolved solid. Fortunately, the suspended solids can be filtered out of the water. The dissolved solids involve more complicated removal processes.

A wastewater treatment plant must break the treatments down into three categories. The primary, secondary, or advanced treatment will depend on the level of purification that is desired. The primary stage will remove pollutants that will settle, float, or are too large to pass through the screening devices. The secondary stage is required by the Clean Water Act for publicly owned treatment works. The second treatment involves similar techniques as in the first stage, except microbial oxidation of wastes is also required. This means that microorganisms will be used to purify the water much as would be seen in nature. The secondary stage should remove 90% of the suspended solids.

Thus, the importance of making a polyethylene milk jug process a "green" process can strike home when one thinks about the valuable tax dollars that are lost to wastewater treatment. Could the engineer minimize the job of the public works department and the perhaps avoidable job of removing suspended solids from polyethylene waste? The plastics engineer may have to decide which is more important, the earth or the milk jugs made of high-density polyethylene.

Low-Density Produce Bag. This profile is rather similar to the high-density version, with the addition of olefins manufacturing that will contribute to air and water pollution. The polymerization step differs somewhat from the low-density version process in that it produces more atmospheric emissions, mainly nitrogen oxides, hydrocarbons, and sulfur oxides. The polymerization step also accounts for most of the process energy.

ABS Dairy Tub. A complex plastic system, an ABS diary tub production will produce nitrogen oxides and hydrocarbons. The combination of nitrogen oxides and hydrocarbons is dangerous. When the two reactants are mixed with sunlight, pollution occurs in the form of photochemical smog. This smog is responsible for chest constriction and irritation of the mucous membranes of people. It causes rubber products to crack. It also damages vegetation. Thus, the simple idea of making ABS dairy tubs can reduce the quality of our lives and our products [41].

Polystyrene Vending Cup. The manufacturing process of polystyrene products such as a typical cup results in the highest atmospheric emissions of all plastic materials. Polystyrene production will produce hydrocarbons and oxides of nitrogen, and the most waterborne waste to date.

The process of making a polystyrene vending cup can be split into two steps. The first step is expansion. The process is called expandable bead molding. During expansion, the particles that contain the so-called blowing agent must be heated. The increase in temperature causes the polymer to soften and the blowing agents to vaporize. The second step is molding. During this process, the expanded particles are blown into molds with a stream of air from a fill gun. In order to remove the cup from the mold, the parts are stabilized by producing a vacuum. Once the vacuum has been created, the inner wall of the mold are sprayed with water. This results in diffusion of gases from the cells and a reduction in temperature [41].

6.12.2   Impact in Industry

A summary of impacts, just in the manufacture of plastics, to the environment is tabulated in Table 6.2.

Recently in the automobile industry there have been changes in the way auto makers are looking at the plastics' environmental problem, for which they are adopting "green" attitudes [42,43]. In North America, Ford Motor Company is the first to recycle salvaged plastic parts, from old and worn out models, back into newly produced vehicles. For example, by using a material reformulated from salvaged plastic bumpers, new taillight housings for the 1993 Ford Taurus can be made.

In terms of polycarbonate, which is used for bumpers on the Ford Sierra, recycling enables the manufacture of bumper brackets for the Monde. Another major application is the grille opening reinforcements on various Ford trucks and cars. This is being produced using recycled plastic bottles.

As for defective compact disks, Chrysler has used them in the new Eagle Vision and Dodge Intrepid's overhead console base.

Today, the world leader in automotive recycling technology is Germany. The processing of plastics for automobile components is taken very seriously. Since it was discovered that over 85% of all shredded waste can be recovered without any harmful side effects to the environment, Germany has been very active in recycling. Volkswagen, for example, sees that processing the approximately 250 pounds of more than 20 types of plastics typically found in today's automobiles is one of the biggest challenges of recycling. Currently German-built cars, such as the Opel Calibra, contain fenders and liners made from spent battery cases, production scraps, and bumpers. Opel will soon manufacture soundproofing, floor mats, and air cleaner housings from recycled ABS and polypropylene.

For automobile interiors, recycled polypropylene performance has made it possible to manufacture "green" plastic structures. Some examples are instrument panels, car seats, door panels, control consoles, and floor covering.

Taking the previous examples into account, industry seems to be heading in the right direction when it comes to environmental concerns involving plastics. With the aid of recycling used plastics and the use of biodegradable plastics, the plastics

Table 6.2. Total Impacts for 1 Million Containers of Each System

	Raw Materials (pounds)	Energy Mill. BTU	Water thou. Gals.	Solid wastes Cubic Feet	Atmos- pheric Emissions (lbs)	Water- borne waste Pounds	Post Cons. Waste
PVC	200,426	12,177	2,007	965	57,365	3.914	5,317
HDPE	8,712	7,515	726	306	27,385	4,081	3,175
LDPE	834	540	44	21	1,983	248	194
ABS	1,631	1,928	491	75	6,892	1,135	706
HIPS	577	550	215	13	1,689	418	226

industry may earn a better reputation with the environmental community. It is noted, however, that plastics are here to stay. Hopefully, with new techniques, and ones that are currently being developed, there will be more and more ways to recycle and reuse discarded plastics.

6.13   PLASTICS RECYCLING

6.13.1   Resin Recyclability

Recycling would seem to be an easy way to reduce municipal waste, keep the environment clean, and save money. Unfortunately, plastic recycling can be very expensive, and can require a lot of complicated processing to produce a product with reduced qualities. It is for these reasons that many companies try to avoid recycling and the use of recycled plastics in their products. Recycling technology, however, is improving and becoming more economically viable. In fact, as the cost of recycling drops and the price of oil increases, plastic waste will become valuable. The public is also demanding more and more recycled goods from companies. The public demand is making recycling more appealing.

About 20% by volume of municipal waste is plastic material [44]. Most plastics are highly resistant to degradation, which means that most of the plastic refuse in dumps today will stay there forever unless it is recycled. The volume of discarded plastics is ever increasing as more and more companies are using lightweight plastic in their durable products as a substitute for other heavier materials and packaging. Plastic packaging is an especially big concern because the life cycle averages less than a year. Efforts to make degradable plastics that would decompose in landfills have been futile. Landfill waste materials lack enough light, air, moisture, and nutrients to sustain microbial action. There are also problems with controlling the triggering of decomposition so that the product does not start to decompose before its service life has ended [45].

The process of recycling consists of collecting, sorting, reclaiming raw materials, and marketing the materials. Sorting consists of the separation of plastics by plastics type.   This is especially important because most plastics are not compatible. Often the mixture of different plastics will yield a product with very poor characteristics. The only way to recycle a plastic back into its pure form is to remove any and all contaminates. This task is the most difficult and expensive. Ideally, each plastic would be separated on a macro scale while the materials are still in large pieces. This is not easily done because often single pieces can contain more than one plastic type. Separation of connecting pieces can be difficult and imprecise, and it is often difficult to recognize each of the numerous plastic types. Most products made out of plastic contain pigments and other additives to alter physical properties and appearance, which are impossible to remove physically. As a result it is often more economical to recycle commingled plastics that are created from a mixture of plastics. The mixture can either be purified by molecular separation, called primary recycling, or made into a refined commingled plastic. This process requires less sorting and processing, and is more economical. The drawback, however, is that it produces a product of reduced quality. This type of recycling is known as secondary recycling. In tertiary recycling the plastic waste is reduced into basic chemicals. The last recycling option is called quaternary recycling and involves extracting energy from the waste plastic by incineration [46,47]. Quaternary recycling is a reasonable option when the scrap plastic is too difficult to recycle by other means because of the plastic type, or its chemical makeup has changed excessively from virgin due to previous recycling or other processes. Table 6.3 can serve as a quick reference of plastics recyclability [48].

Primary Recycling. Primary recycling involves converting plastic waste into products with characteristics similar to the original product. This, of course, is the most ideal form of recycling. It is also the most difficult to achieve. This is due to the difficulties in recycling contaminated (containing dirt, labels, etc.) or mixed (more than one type of plastic) waste streams, which are usually the available feedstock. Chemical contamination is a concern with food packaging since there is the possibility of interaction between the food product and the contaminates. Contaminates can also adversely affect the appearance of some plastics, like translucent milk bottles. The altered appearance can make consumers wary of the quality of the product.

Primary recycling, however, can be a viable option if the waste can be separated for a single resin. Industry has been recycling its own unmixed waste streams of scrap and trim plastics from assembly processes for many years. Clear, unpigmented resins are best for recycling because they give the user the widest range of application choices. Collection programs such as bottle deposits have created an unmixed waste stream for recycling. Because of this, HDPE and PET compose 60% of all recycled plastics. There is a minimum premium, however, of about 8–10 cents per pound for the recycled plastic. The lack of separate collections is the major reason why plastics are not recycled as much. There are means of recycling mixed plastic streams in the primary sense, though; Polysource Mid-Atlantic has developed a sophisticated hydrocyclone to process the entire postconsumer plastic waste stream and refine each type of plastic waste into a virgin resin substitute. This does not require a feedstock of only plastic waste.

Secondary Recycling. Secondary recycling involves producing products with less demanding physical and chemical characteristics. There has been a lot of work done in this area. Secondary recycling has been the usual method of recycling mixed plastic waste to produce plastic lumber and other bulky objects that once were constituted of wood. There have been many developments made to design melting processes to avoid the production of harmful chlorine and cyanide gases from polyvinyl chloride and polyethylenes. Thermosets have also been recycled in this way, by grinding them into a fine powder and using it as a filler for virgin resins. In general, secondary recycling is only a temporary solution. This is because the recycled products do not replace the products made from virgin resins. The degradation of the bonds in the plastic after recycling will also eventually force the resin into a tertiary or quaternary recycling operation.

Table 6.3. Plastic Recycling Reference Guide

| Plastic | Recycle | Monomer | | Polymer | |
		Raw Material	Waste	Effici-ency%	Waste
PET	**1** -15-20% recycled 8-10 ¢ premium **4** -20050 btu/lb	Ethane Propane		10-30	Methane, hydrogen Graphite
HDPE	**1** -1 % recycled 8-10 ¢ premium **4** -20000 btu/lb	Ethane, Propane		10-30	Methane, hydrogen Graphite
PP	**2** -gas free melting process **3** -Pyrolysis **4** -20030 btu/lb	Petroleum products - natural gas or light oil		80-85 12000-30000 g PP/g cat	Reaction gasses recycled
PS	**1** -1% recycled **3** -complete pyrolysis at 500 C **4** -17870 btu/lb	a) Benzene, Ethylene b) Styrene	51% Benzene & 8% polyethyl benzene recycled 22% aromatic hydrocarbon	30-35	
PVC	**1** -3% recycled **2** -Gas free melting **3** -complete pyrolysis at 500 C **4** -27200 btu/lb	Ethylene & Benzene	Ethylene dichloride HCl recycled	85-90	for 50 parts monomer 0.01 gelatine 0.1 trichloro ethylene 0.01 caproyl peroxide 90 demineralized water

Note. **1** = primary recycling; **2** = secondary recycling; **3** = tertiary recycling; **4** = quaternary recycling; a, b: process steps

Tertiary Recycling. Tertiary recycling involves the production of basic chemicals and fuels from plastic waste polymers [49]. There are two main types of tertiary recycling. These are pyrolysis and hydrolysis. Pyrolysis involves heating the plastic waste without oxygen to drive off volatile compounds, leaving behind carbon and ash. The products include combustible gases (chemical feedstocks or purifiable to pipeline quality) and liquid products that can be used as a low-sulfur fuel oil [50]. The exact makeup of each operation depends on the composition of the waste stream, which can vary significantly. Hydrolysis involves subjecting plastic waste to superheated steam for several minutes. This breaks down the polymer's chemical structure to produce the basic chemicals for new resins. This method produces a more uniform product, but requires that the input be uncontaminated and of a single resin type. Tertiary recycling is usually used to recycle manufacturing waste plastic. Some of the advantages of this type of recycling include:

- It can accommodate contaminated waste streams.
- It is less polluting than incineration.
- It is self-sustaining (more energy is produced than is consumed in the process).
- The volume of waste can be reduced by up to 90%.

Recently, there has been an expressed interest in expanding tertiary recycling from an industrial process to a more commercial one. The attraction has been that all of the tedious collection and separation steps for primary or secondary recycling are avoided. Estimates have been made that tertiary recycling could be used commercially to generate 80 million barrels of oil a year, if it were applied to the municipal waste stream.

Quaternary Recycling. The last form of recycling, quaternary recycling, involves the retrieval of the energy content from burning plastic waste. While this is the least attractive method, it is useful for recycling plastics that are unable to be recycled by any other means. These plastics also have a rather large heating value. The heating value is much larger than that of coal (average heating value for plastic is 1800 Btu/lb, vs. 1300 Btu/lb for coal), so there is a good deal of energy that can be recovered. In fact, most of the fuel value of the original oil used to make the plastic can be retrieved. The disadvantage, of course, is that toxic fumes can be produced in the combustion process. However, when an incineration plant is operated at the proper temperature and under the proper conditions, the dioxin levels can be even lower than those produced by most home heating systems.

6.13.2   Incineration of Plastics

Resource recovery or incineration is the most viable disposal method besides landfilling and recycling. This process of burning waste (to generate steam to power turbines and produce electricity) is in use today by many cities throughout the world. A by-product of incineration, however, can be toxic air pollution.

The incineration process requires an analysis of the quantities of elemental components present in the plastic. This is done in order to determine the products of combustion, combustion air requirement, and the nature of the off-gas or combustion products. During this procedure, the following element percentages are normally determined:

- carbon
- hydrogen
- sulfur
- oxygen
- nitrogen
- halogens (chlorine, fluorine, etc.)
- heavy metals (mercury, lead etc.)

Additionally waste moisture content, heating value, and inorganic salts must be considered. The waste moisture content is important because the greater the moisture content, the more fuel is required to destroy the waste. The heating value is important because in order to facilitate combustion, the material must lend Btu's to the combustion process. Normally a waste with a heating value less than 1000 Btu/lb is not applicable for incineration.

Inorganic salts (alkaline salts) are difficult to dispose of in a conventional incineration system. A significant fraction of the salt will become airborne. It will usually collect on furnace surfaces, creating a slag that severely reduces the ability of an incinerator to function properly. High sulfur or halogen content normally results in the generation of acid-forming compounds in the exhaust gases. Overall, the quality of the plastic waste plays an integral part in the incineration process.

A major concern with incineration is the safety of burning increasing amounts of plastic packaging containing a large mixture of different polymers. Some resins, particularly PVC and chlorinated polymers, contribute to the formation of corrosive hydrochloric acid gases. Additives in these polymers are released during combustion. The organic additives are burned, but inorganic additives, such as cadmium sulfide, are left in the bottom ash of the incinerator. Both fly ash and bottom ash must then be disposed of as hazardous waste due to concentrations of heavy metals.

Typically, the plastics are burned in a mass burn system. The combined plastic waste is brought to a storage pit or tipping floor where a conveyor system transfers the plastic to a hopper. The waste is then shredded and fed into the furnace. Once the waste enters the furnace, a combustion air system and auxiliary burners aid in the complete combustion. Air pollutant equipment, ash quench, and removal systems are then used to filter the emissions as they rise up the stack of the incinerator. These systems ensure a clean discharge into the surrounding environment, and therefore it is crucial that they are constantly monitored and maintained.

In 1993, Industronics, Inc., performed a test burn on PET bottles in a pilot waste-to-energy system located in South Windsor, Connecticut. The company baled whole PET bottles and incinerated them. The test resulted in a total volume reduction of 99.0% and a weight reduction of 99.6%. Industronics, Inc., found PET to be an excellent source of energy with a rating of 12,000–15,000 Btu/lb: twice that of coal. The particulate emissions met EPA standards, no acidic gases were formed, and organic constituents were completely destroyed.

6.13.3 Health Concerns

Many U.S. cities are eyeing incinerators as they find themselves increasingly turned away from landfill sites. Americans toss out at least 150 million tons of trash every year, and 90% of that winds up buried. An April 1987 Worldwatch Institute Study of the world's growing garbage glut revealed that by 1990 half the cities in the United States will have exhausted their landfills. As reports of groundwater contamination from buried garbage grow, city officials across the nation are encountering heavy local opposition to the opening of existing landfills and to the opening of new ones. Cities often must resort to trucking their trash to rural areas, or even to other states [51].

The issue of incinerator ash—in particular, how to dispose of it—has drifted to the forefront of the burn-plant debate. In EPA tests, every sample of fly ash, the fine particulate matter trapped in the plants' air pollution control devices, showed unacceptable levels of toxic metals such as lead and cadmium. Tests of bottom ash, the unburned residue that collects on an incinerator's grate, showed unacceptable levels of these elements in 10–30% of the test cases. Concentrations of the potent carcinogens dioxin and furan are also present in fly ash.

In the long run recycling is also cheaper than either dumping or burning, advocates add. The 1-ton bales of rotting paper on St. Pierre's barge, for example, might have fetched up to $20 a ton from recycles but would have cost at least $40 a ton to dump—if a landfill had been willing to accept them. Cities should burn trash only as a last resort to their garbage crisis, activists say—and only after less damaging waste reduction, recycling, and source-separation programs have been implemented.

A special report examines efforts to create environmentally sound chemical. Articles discuss the changes in the marketplace being caused by the focus on "green chemicals," the solvent industry's move toward water-based systems, the replacement of chlorine with oxygen products in the paper industry, degradable plastics products, the biological pesticides industry, and political and health concerns related to oxygenates.

A main health concern of incineration in the release of heavy metals into our environment. Heavy metals are chemical elements that cannot be destroyed by incineration [52]. The EPA indicates that batteries and plastics are a major contributor of both lead and cadmium. Because of their permanent nature, heavy metals are accumulated both in the environment and within human bodies. Thus, long-term exposure, no matter how low the level, will produce substantial heavy metal levels in humans and the environment.

Metals and dioxins, which are a by-product of incineration and are found in the fly ash, can be minimized through careful control of the incineration process. Otherwise, they can be emitted and will contaminate the surrounding area. The chemicals will fall on vegetation, leak into the groundwater, and thereby enter into the food chain. This will lead to ingestion of these chemicals by humans. This will elevate the cancer risk of the exposed individual.

Common dioxins released through incineration are polychlorodibenzo-*p*-dioxins (PCDD) and polychlorodibenzofurans (PCDF). These two dioxins are considered carcinogens, and the danger is long exposure to low levels in the surrounding environment, as well as the induction of these chemicals into the food chain. The exposure to these carcinogens has an additive effect.

The risks associated with incineration by-products must be taken seriously. Although risk through inhalation is low, the long- term effect of low-level emissions into surrounding soils and vegetation are not yet fully understood. These routes of contamination will have negative long-term effects on the environment. Currently, there is uncertainty as to the extent of the risk, but sufficient information is available on the concentration and bioavailability of organic and inorganic compounds in both air emissions and ash residues to indicate the reality of these risks.

6.14   SUMMARY

By providing a detailed description of various forms of plastics, and how they are utilized and strengthened with reinforcers, it is easier to understand how complex environmentally conscious engineering principles are used with plastics. There is a major challenge for plastics manufacturers, product manufacturers that use plastics, and plastics consumers to be aware of the environmental impacts if we do not begin to recycle and develop new ways of utilizing recycled plastics.

By providing proper recycling measures to plastics, namely, polyethylene, PVC, polycarbonate, polypropylene, and ABS, waste reduction will be extremely successful, resulting in a cleaner environment while still retaining the benefits that plastics have provided us with.

6.15   BIBLIOGRAPHY

1.   Bikales, N.L. *Mechanical Properties of Polymers*, John Wiley and Sons, New York, 1971.

2.   Boettcher, F.P. *Environmental Compatibility of Polymers*, Emerging Technologies in Plastics Recycling, Symposium Series 513, ed. G.D. Andrews. American Chemical Society, Philadelphia, 1992.

3.   Brennan, A.B. Surface Modification of Polyethylene Fibers for Enhanced Performance in Composites, *Trends in Polymer Science*, pp. 12–15, January 1995.

4.   Briston, J.H., and Gosselin. *Introduction to Plastics*, Butterworth and Co., 1970.

5.   Brydson, J.A. *Plastics Materials*, 4th ed., Butterworth Scientific, Boston, 1982.

6.   Hanco, N.L., and Mayer, R.M. *Design Data for Reinforced Plastics: A Guide for Engineers and Designers*, Chapmen and Hall, New York, 1994.

7.   Katz, H.S., and Milewski, J.V. *Handbook of Fillers and Reinforcements for Plastics*, Van Nostrand Reinhold, New York, 1978.

8.   Resins, *Encyclopedia Britannica*, 1970.

9.   The Software Toolworks Series, "Multimedia Encyclopedia", Version 1 PB, 1992.

10. MacDermott, C.P. *Selecting Thermoplastics for Engineering Applications*, Marcel Dekker, New York, 1984.
11. Polyvinyl Chloride, *Chemical Week*, April 5, 1995.
12. Global Markets for Chlorine and PVC: Potential Impact of Political Opposition, *Plastics Engineering*, August 1994.
13. Callister, W.D. *Materials Science and Engineering, An Introduction*, 2nd ed., John Wiley & Sons, New York, 1991.
14. May, W.P., and Hunter, D.R. Polyvinyl and Vinylcopolymers, *Modern Plastics Encyclopedia*, 1986–1987.
15. Thompson, T., and Klemchuk, P. *Stabilization of Recycled Plastics*, Emerging Technologies in Plastics Recycling, ACS Symposium Series 513, ed. G.D. Andrews, American Chemical Society, Philadelphia, PA, 1992.
16. Guillet, J. *Polymers with Conditional Lifetimes*, Polymers Science and Technology, Vol. 3, ed. James Guillet, Plenum Press, New York, 1973.
17. English, L. K. Some Common Problems with Thermoplastics, *Materials Engineering*, August 1989.
18. Masia, L. *The Role of Additives in Plastics*, Wiley and Sons, New York, 1974.
19. Miller, B. Ressing to Turn to When the Heat Is On, *Plastics World*, 48, pp. 40–45, August 1990.
20. Hurter, B. Nikon F4, *Petersen's Photographic Magazine*, 17, 41–45, March 1989.
21. Avery, S. Plastics Bounce Back, *Purchasing*, 117, pp. 73–74, October 1994.
22. Thai A300-600s Delivered with Updated Cockpit Styling, *Aviation Week & Space Technology*, 123, p. 89, November 1985.
23. Smocl, D., Blends Will Fuel Growth at GE Plastics, *Plastics World*, 50, pp. 19, September 1992.
24. Curlee, T. R., Targets of Opportunities for Plastics Recycling and Source Reduction, *Plastics Recycling as a Future Business Opportunity*, pp. 21–45, Techmonic, Lancaster, PA, 1990.
25. Zhenxing, L. Look at China's Plastics Market, *Plastics World*, 50, pp. 58–59, November 1992.
26. Kocsis, S., and Karger, J. *Polypropylene, Structure, Blends, and Composites*, Vol. 1, Chapman and Hall, 1995.
27. Ficker, H.K. Has Phase Process for PP Surpass Other Methods, *Plastics Engineering*, pp. 29-32, February 1987.
28. *Modern Plastics Encyclopedia*, Vol. 63, Number 10A, McGraw-Hill, New York, October 1986.
29. Arimond, J. "Creep Characterization of Phenolic Composites for Fastening and Sealing Design," SAE Technical Paper Series. Bulletin No. 931027, March 1993.
30. Polycarbonate Research Focuses On Flow, Alloys, and Recycling, *Research and Development Magazine*, July 1994.
31. Kirkland, C. Non-opening Mold Hikes Productivity, *Plastics World*, 51, p.

59, March 1993.
32.   Coesel, A. Recycling Second Tier Plastics, *Chemical Marketing Reporter*, p. 25, June 28, 1993.
33.   Rennie, C. The Pitfalls and Promises of Plastics Recycling, *Plastics Recycling as a Future Business Opportunity*, pp. 56-66, Techmonic, Lancaster, PA, 1990.
34.   Cross, J. A., Welch, R., and Hunt, R. G. "Plastics: Resource and Environmental Profile Analysis", Midwest Research Institute, Kansas City, MO, 1974.
35.   Erwin, L., and Hall, H., Jr. "Packaging and Solid Waste Management Strategies," A Management Briefing for the American Management Association Packaging Council, 1990.
36.   Naj, A.K. Forget Recycling, *The Wall Street Journal*, B1, September 29, 1993.
37.   Wolf, E., and Feldmen, E. *Plastic: American's Packaging Dilemma*, Island Press, Washington, DC, 1991.
38.   Organd, J. Biodegradable Polymers Crop Up All Over Again, *Plastics Technology*, p. 60, August 1992.
39.   Palmisano, A.C. and Pettigrew, C.A. Biodegradability of Plastics, *Materials Engineering*, August 1989.
40.   Potts, J.E. *Biodegradabilty of Synthetic Polymers*, Polymer Science and Technolgy, Vol. 3, ed. James Guillet, Plenum Press, New York, 1973.
41.   Masters, G.M. *Introduction To Environmental Engineering and Science*, Prentice Hall, Englewood Cliffs, NJ, 1991.
42.   Fleming, A. Green Horn,*Automotive News Insight*, April 20, 1993.
43.   Siuru, B. From Scrap Heap to Showroom, *Mechnical Engineering*, November 1990.
44.   Mustafa, N. *Plastics Waste and Mangement*. Marcel Dekker, New York, 1993.
45.   Andrews, G.D., and Subramanian, P.M. *Emerging Technologies in Plastics Recycling*, American Chemical Society, 1992.
46.   Brunner, C.R. *Incineration Systems: Selection and Design*, Van Nostrand Reinhold, New York, 1984.
47.   Denison, R., and Ruston, J., *Recycling and Incineration: Evaluating the Choices*, Island Press, Washington, DC, 1990.
48.   Ayles, W. Phenolics: Heat Resistance, Stability and Electrical Properties Spur New Applications, *Modern Plastics*, pp. 88-90, November 1993.
49.   New Technology Fuels 261 (percent) Rotomolding Gain in '90, *Plastics World*, 49, pp. 71–73, May 1991.
50.   Fire, F.L. *Combustibility of Plastics*, Van Nostrand Reinhold, New York, 1991.
51.   Arzoumandis, G. and Karayannis, N., Fine Tuning PP, *Chemtech*, pp. 43–48, July 1993.
52.   Hattemer, H.A., and Travis, C. *Health Effects of Municipal Waste Incineration*, CRC Press, Boston, 1991.

53.   The Guardian Magazine. A Controversial Theory on Crib Death,*World Press Review*, July 1995.

54.   Grandilli, P.A. *Technician's Handbook of Plastics*, Litton Educational Publishing, Inc., New York, 1981.

55.   Katz, H.S. *Plastics: Designs and Materials*, Macmillan, New York, 1978.

56.   Schut, J.H. A Plastic Odyssey from Button to Abrasive to Bathroom Sink, *Plastics World*, 52, pp. 8, September 1994.

6.16   PROBLEMS / CASE STUDIES

6.1   PVC Is it Safe for Use?: The primary purpose of this case study is simply to expand the mind and explore another aspect of the plastics world.

The *World Press Review* of July 1995 [53] told a very interesting story about a discovery involving SIDS (Sudden Infant Death Syndrome or crib death) with PVC. This medical phenomenon was formally described by A. M. Barrett in 1953. He explained that crib deaths had significantly increased during the past few years. These sudden unexplained deaths occurred in infants mainly during sleep. Unfortunately, the situation got worse. In 1986, 1500 babies in England and Wales alone had died of this terrible yet unexplained phenomena. Deaths occurred in the United States, Australia, and other countries, but not in China, India, or Japan.

An answer to this puzzling problem began to form at a wedding in England. A man named Barry Richardson had rented a tent for his daughter's wedding. The man who rented the tent to Richardson had heard that the manufacturers were having problems with the deterioration of the PVC-reinforced fabrics used in tents. Richardson was a professional with a lot of experience in the field of material deterioration and its associated health risks. His simple answer was that the deterioration was due to fungi. After a series of events, the word that the fungi was the cause of tent deterioration reached the tent manufacturers. The tent makers did exactly what we may have expected. They added an antifungal additive to the PVC-reinforced tents. Remember that plastic compatibility is always an experimental process. This experiment proved to be a failure. When this antifungal additive is introduced in large quantities, the active fungal deterioration can convert the fungicide to toxic arsine gas.

a.   What do all of these Englishmen in tents have to do with SIDS? Take the tent example and link it to SIDS.

b.   One case shows a family who went to visit a friend. The family, along with their baby, decided to stay the night. The host decided that the baby should sleep in their son's crib. The next day the friend's baby had died in the night. What could have caused the friend's baby to die while the host's baby had not?

6.2   High-Performance Plastics in the Automotive Industry: When phenol is reacted with formaldehyde a special kind of plastic known as phenolic resin is produced. This material has been classified as a heat-cured themoset. The cross-linked structure formed by this reaction results in a material that is heat resistant, dimensionally stable, creep resistant, and may be fabricated into components with tight dimensional tolerances. Phenolics are rarely used without some sort of reinforcement systems. These systems may include wood flour (refined sawdust), mineral, cotton flock, chopped fabric, short or long glass fibers, Teflon, and nylon. Applications for phenolic materials include automotive, appliances, and electrical components. These versatile compounds are the primary choice of automotive designers when optimizing cost, performance, and dependability where applications involving elevated temperatures under load are required.

Chemically, there are two types of phenolic resins; Resole, or those that require a single-stage production process, and *Novolac*, which require a two-stage production process. The Novolac type resin is much easier to control in the manufacturing process than the resole type. Consequently, Novolac resins are predominately found in industrial applications [48].

Dry granular phenolic molding compounds are divided into two broad classes: general-purpose and high-performance engineering grades. The general-purpose wood-flour-filled phenolics are extensively used in knobs, handles, and bases for many common appliances. They also are widely used in high-temperature applications such as ovens or wiring devices. The mineral- and glass-filled compounds are typically used when high strength, hardness, wear resistance, electrical properties, impact strength, or high temperature characteristic is required. Furthermore, high-performance grades have found many applications in the automotive industry.

a.   Estimate the number of the vehicles that are currently being scrapped per year and the percentage of material by weight that is recycled. What is the percentage of plastics recycled?

b.   Select a few parts of the vehicle that are made of plastic (e.g., instrument panels) and estimate the number of plastics components being used in their manufacturing.

c.   Name at least six areas where technology can improve plastics recycling in the automotive industry.

d.   Due to the continuing demands on automakers for higher performance, more fuel-efficient vehicles, great emphasis has been placed on weight reduction programs. Would plastics be considered as a good choice in these programs? What parts of the vehicle could be replaced by plastics?

6.3   Residential Recycling: Recycling works only when people participate. Back in the early to mid 1980s the average citizen was given the opportunity to recycle some refuse at buy-back and recycling drop-off facilities. Only about 5% of the population went through the trouble of sorting, separating, and cleaning their recyclables and taking them to a recycling center. Those who went to buy-back facilities were motivated by money, while the others were motivated by a sense of environmental and civic responsibility. By the mid 1980s beverage container buy-back programs had been started in nine states. It was the first large-scale collection of polyethylene terephthalate (PET). The mid 1980s also marked the start of curbside residential recycling programs in some communities. Most communities did not include plastics recycling in their programs because of its low value and difficulty to recycle satisfactorily. By the end of the 1980s plastic, chemical, and packaging companies had invested a lot of effort in promoting and developing methods to make plastics recycling easier. By the beginning of the 1990s the effort paid off with 75% of curbside recycling programs including plastics. The success of curbside recycling is owed mostly to the recycling container. It was found that communities that distributed a recycling container to each residence had twice as much participation in their recycling programs than communities without containers. The container provides residents with a place to store recycleables, and it acts as a reminder. It is also a form of peer pressure as each resident feels compelled to have a full container curbside on collection day for all his neighbors to see. Participation levels also rose as the frequency of pick-ups and the convenience of the service increased. Frequent pick-ups meant residents had to recycle every day to have a full container on pick-up day. They were not able to forget about recycling for a while between pick-ups, because the pick-ups were so frequent. And when the service is convenient people are happier to participate. Participation also grew as a result of promotion, education, and awareness. These days there are more and more news reports and articles concerning environmental issues. As people become more aware of environmental concerns they look to participate in change. Recycling is one way a person can participate in helping the environment. As plastics recycling processes become better and more efficient the value of waste plastic, for both consumers and industry, will rise. What was once considered garbage will become a valuable resource.

a.   Evaluate a number of parameters for this process, such as materials, waste produced, etc.

b.   What would be the necessary steps that manufacturers of plastic materials should consider for recycling of their products?

6.4   ABS is a very common material in automobiles. It is typically specified in both large and small components in the interior, and less frequently for the exterior and under-hood areas. Based on Section 6.7 and Table 6.2 cite at least five reasons for these choices.

6.5   Consider the redesign of a chain saw (or any of your garden tools) from many metal parts to a small number of mostly plastic parts. Consider pros and cons of the two designs from an economic and an environmental viewpoint; for example, plastic is cheaper but probably more difficult to recycle, but then metals require coatings, such as plating or painting, that generate pollutants: toxic heavy metal solution residues or volatile organic solvents. Characterize as many steps of these processes relative to both cost and environmental impact. Develop a comparison scheme that allows for weighing the pros and cons of these "apples verses oranges" type choices. Are these comparisons likely to be characteristic for general cases of plastic versus metal parts? Summarize a set of rules that can be used for general design guidelines.

6.6   Polyethylene comes in several grades or densities. Name an item that you know that could be made from each of these grades.

6.7   What is the difference between a thermoplastic and a thermosetting plastic? What is it about thermosets that makes them so difficult to recycle?

6.8   What are some of the reasons why most plastic waste cannot be recycled back into its original virgin form?

6.9   A Semester-Long Class Experiment.   Materials needed:
        2 paper grocery bags   2 Styrofoam cups
        2 plastic grocery bags   1 shovel
        1 fine-tuned measuring device (Vernier caliper/micrometer)

Obtain these materials. Take one paper bag, one cup, and one plastic bag and bury them under 1 foot of soil. Take the other set of items and place them in a sunny window. Periodically dig up the items and note the degradation of the materials. Measuring may be one possible way to note the changes. A simple journal log of visual notations may be more appropriate. Note the changes in the cup and bags that were placed in the sun as well.

Encourage the class to consider the implications of the changes. If no changes occur in the time that the class has for the experiment, this can be a significant statement about how "green" these day-to-day items are.

APPENDIX 6A. POLYMER PROPERTIES

Polymer	Corr. Water, Chem-Resist.	Tensile Strength (Kpsi)	Impact Strength (ft-lb/in)	Type, Class & Group	Typ. Applic.	LOI	Igni-tion Temp	Heat of Combust (Btu/lb)
Acetals	good	10.0	1.4	TP/CRY B	Bushings Aerosols	14.7-16.0	825	7300
Acrylics	excell	6.0	0.3	TP/CRY B	Lights Blister packs	17.0-19.4	806-1040	14000
A.B.S.	good	6.8	10.0	TP/CRY C	Bath tubs / Canoes	18.0-39.0	780-915	15500
Cellulose	excell	5.0	5.0	TP/CRY B	Audio tape & tooth brushes	18.0-27.0	280-1004	10100
Cellulose Acetate	excell	6.0	6.0	TP/CRY B	Combs Blister packs	18.0-27.0	887	10100
Cellulose Nitrate	excell	6.0	5.0	TP/CRY B	Pens Cigarette Filters	18.0-27.0	286	10100
Cellulose Triacet.	excell	6.0	5.0	TP/CRY B	Face Shields Buttons	18.0-27.0	1004	10100
Ethyl Cellulose	excell	7.0	0.75	TP/CRY B	Light cases Steering wheels	18.0-27.0	565	10100
Polytetra Fluoro Ethylene	excell	4.0	2.5	Powder Process/ AMO/D	Non-slip cookware Pipes Bearings	95	986-1076	2000

APPENDIX 6A. Continued

Polymer	Corr. Water, Chem- Resist.	Tensile Strength (Kpsi)	Impact Strength (ft-lb/in)	Type, Class & Group	Typ. Applic.	LOI	Igni- tion Temp	Heat of Combust (Btu/lb)
Polyvinyl Fluoride	excell	8.0	2.5	TP/AMO D	Laminate	22.6	1000	9000
Poly Vinyliden Fluoride	excell	12.0	3.0	TP/AMO D	Chem. piping Rods	43.7	980	4000
Poly Butad.	good	10.0	6.0	TP/CRY A	Syn. rubber Tires	18.3	860	20000
Poly Arylates	good	10.0	0.35	TP/CRY A	Wire Insulat. Tints	15.0- 25.0	825	10000
Chlorin. PVC	good	16.0	10.0	TP/CRY D	Piping Skylight Frames	45.0- 49.0	1035	8500
Nylons	excell	20.0	10.0	TP/CRY C	Engine fans	20.0- 30.0	795- 990	14000- 16000
Poly acrylo nitrile	good	12.0	0.4	TP/CRY C	Blister packs Meat pckaging	18.2	896	14000
Poly carbonate	good	9.0	2.5	TP/CRY B	Auto lights Compact disks	23.0- 44.0	986- 1076	19000
Poly butylene Tereph thalate	good up to 140 F then poor after	17.0	1.8	TP/CRY A	Auto body Panels Mirrors	22- 46	653- 796	8000

APPENDIX 6A. Continued

Polymer	Corr. Water, Chem-Resist.	Tensile Strength (Kpsi)	Impact Strength (ft-lb/in)	Type, Class & Group	Typ. Applic.	LOI	Ignition Temp	Heat of Combust (Btu/lb)
Poly Ethylene tereph thalate	good	23.0	1.9	TP/CRY AMO/B	Mouth wash Bottles Carpet	20.0-22.7	896	10000
Poly Ethylene	good	3.0	1.0	TP/CRY A	Grocery bags Milk bottles	17.4	660	20000
PolyProp ylene	good	9.0	3.0	TP/AMO A	Rigid Package	17.0-28.0	806-824	20000
Poly styrene	good	9.0	0.3	TP/AMO A	Blister packs Egg cartons	18.1	910-1000	18000
Poly Urethane	good	6.0	5.0	TP/TS/ CRY/ AMO/C B	Foot wear Blister packs	17.0-21.4	780	12000
P.V.C.	excell	16.0	0.7	TP/AMO D	Bottles Pipes	Rigid-45 Flex-19-40	Rigid-1035 Flex-850	8500
Polyviny-lidiene Chloride	good	18.0	6.0	TP/AMO D	Saran Wrap	60	986	4300
Acrylic styrene acrylo nitrile	good	8.0	0.35	TP/AMO D	House siding Windows Boat extr	16.6-19.4	900	19000

APPENDIX 6A. Continued

Polymer	Corr. Water, Chem- Resist.	Tensile Strength (Kpsi)	Impact Strength (ft-lb/in)	Type, Class & Group	Typ. Applic.	LOI	Igni- tion Temp	Heat of Combust (Btu/lb)
Stryene acrylo nitrile	good	8.0	0.4	TP/AMO C	Lawn Furniture	18- 28	690	18000
Styrene Butad.	excell	6.0	2.2	TP/AMO C	Shrink wrap Sand. Bags	16- 19	680	20000
Alkyds	excell	12.0	0.4	TS/B	Lenses Watch Crystals	29.0- 63.4	900	6000- 17000
Expoxies	excell	25.0	10.0	TS/B	Flooring Laminant	18.3- 49.0	1100	17000
Phenolic	excell	18.0	3.0	TS/B	Fuse blocks Motor housings	29.0- 66.0	1060	12000
Silicones	excell	14.0	2.0	TS/B	Caulking Sealing Compnds	26.0- 41.0	870	6000
Un Saturated Polyester	excell	20.0	3.5	TS/B	Corvette car body bayliner Boat Hull	22.0- 46.0	990	7700
Urea formalde hydes	excell	16.0	0.3	TS/B	Adhesive coatings	30	1000	7700

Notes. Tensile strength is based on the ASTM D-638 Test (Kpsi). Impact strength is based on the IZOD impact strength test (ASTM D-256). TP = thermoplastic polymer; AMO = amorphous structure (the material shows a second-order transition temperature, at which it changes from brittle to a more elastic form); CRY = crystalline structure (a polymer with a crystal-like structure, exhibiting higher strength and brittleness); GROUP = combustion by-products grouping; LOI = limited oxygen index (limiting concentration of oxygen in the atmosphere for sustained combustion at room temperature; a material with a LOI of more than 21 will not burn in air at room temperature).

PART
TWO

DESIGN EXAMPLES

SEVEN

DESIGN FOR ASSEMBLY AND DISASSEMBLY

Increasing demand for lower cost of goods, due to international competition, has developed a new area of rationalization for increased productivity, which can be achieved by applying new radical methods of production principles. Design for assembly and disassembly (DFA/DFD) are some of these methods that should be considered in the early design stage to cope with the increasing demands of lower cost, higher quality, faster production, improved working environment, and higher wages. When DFA/DFD is implemented with the axiomatic design (AD) theory in an intelligent environment, a new era of design concepts could be realized. This chapter examines some elements of such an intelligent design environment. It provides some methods of design that have to be incorporated to meet the growing pressure from consumers to produce environmentally conscious design and manufacturing practices. The contribution of these methods will therefore be the presentation of fertile areas of research for development of new environmentally friendly products.

7.1   INTRODUCTION

Every year Americans send approximately 180 million tons of municipal solid waste to landfills. This correlates to 4.5 pounds a day for each man, women, and child in the country. Many of today's largest U.S. companies have recognized this as a problem and have begun to design products with the environment in mind from the start [1]. They are choosing materials that are easier to recycle, assemble, and disassemble later. For example, in February 1994, General Motors, Chrysler, and Ford formed the Vehicle Recycling Partnership (VRP) to develop ways to recycle, recover, and reuse as much of the fluff and metal scrap from motor vehicles as possible [2].

One of the biggest problems of recycling postconsumer products is the task of separating the different integral parts of a product or mechanism. Until this era of recycling, products were designed to stay together indefinitely. In the past, designing for assembly (DFA) simply meant putting a product together as fast as possible, while using the least amount of parts. Also, this type of design was concerned with making sure the final product did not come apart. Today, governments and consumers are asking industry to design a rugged product that can be economically taken apart and recycled or reused.

The term recycling, as discussed earlier, means postconsumer products are broken down into a material state similar to their respective raw materials. For example, used metal is broken down into ingots, while plastics are transformed back to pellet form. Although products that use recycled parts are "green," a lot of their value and/or energy that went into making them is lost. Worse yet, these recycled products can be more expensive than virgin material, making them useless. In order for the use of a recycled product to be economical, it needs to be designed for assembly and disassembly. Products can be designed to be used in another product in the form of raw material or in a shape that is more valuable.

7.2   DESIGN FOR ASSEMBLY

Many large companies in the world have found that up to 75% of a product's manufacturing costs are determined before their manufacturing department becomes involved with new product design [3]. By applying design for assembly (DFA) concepts that integrate both design and manufacturing functions and people in one group, tremendous success can be achieved in the transition of research and development to manufacturing. This success is accompanied by involving wide range of people: product managers, engineers throughout different departments, salespeople, supervisors of manufacturing, vendors, and the most important factor, the hourly wage people. When DFA approach is appropriately used the following benefits can be realized:

- Optimization of manufacturing process from the beginning.
- Proper use of available manufacturing and inspection environment.
- Manufacturable and cost-effective product design.
- Interdisciplinary creativity.

The major goal of DFA is concerned with reducing the cost of a product by simplifying its design. This can be accomplished during the conceptual stage of design (at initial attempt of design stage) or downstream of the design stage (to improve existing design). The important criteria is to reduce the number of individual parts to be assembled and to ensure that all parts are easy to manufacture and to assemble. In many current practices, assembly processes are not studied in advance, and later during the manufacturing stage, high cost results due to insufficient ease of assembly.

DFA provides guidelines for better design of a product by reducing the number of components and avoiding tedious movements of assembly motions such as lifting or turning. It also simplifies inventory and record keeping, improves material and production flow, reduces the number of parts in a product, replaces certain parts with better alternatives, and improves the assembly process.

Boothroyd [4] has pointed out this costly problem and developed a new area of DFA with special emphasis on reducing number of components. Even though his classification of assembly methods is divided into three categories depending on production volume (manual assembly, high-speed automatic assembly, and robotics assembly), all of them strive for one goal, reducing the total number of components, which in turn reduces the assembly cost. For many products, comparing the assembly cost is a good way of analyzing product design efficiencies since this is a major criterion for all industrial organizations.

A product design can be evaluated using two criteria. The first is by examining product components and searching for elimination or combination with other components. During this search, production feasibility should not affect the idea generation since this will discourage some possible search. Components that can be combined but are hard to produce can be reanalyzed at a later stage for alternate production methods. The second criterion is by estimating the time of assembly (picking, moving, inserting, positioning, etc.). This should be done with an ultimate goal of comparing different product designs. Eliminating or combining components does not always bring advantageous effect, but resulting components can lead to "impossible to manufacture" components or unusually high production or assembly cost. The following steps are used in order to generate the DFA evaluation of product design efficiency [4]:

- Break the product into a number of components and organize them by list of component names. For each component, determine if it needs orientation during the handling process. Also study how it is fastened to the existing components.
- For each component, determine whether it belongs to a rectangular envelope or a cylindrical envelope. Determine the dimensional and handling requirements (grasp, manipulate, etc.) for corresponding envelopes. Also determine if it needs more orientation before it gets combined with (assembled to) the existing components.
- Reassemble all components and study the assembly processes. Determine what kinds of assembly processes (insertion or other joining) are required.

7.3   DESIGN FOR DISASSEMBLY

In discussing design for assembly and disassembly, it is important to focus on the design problem itself. One of the most complex problems with designing for disassembly (DFD) is the fact that one must first design for assembly. The problem with this is that designing for assembly and disassembly is hard to justify economically. This means that design for assembly and disassembly is often more expensive than dumping the extraneous part in a landfill. Other problems that could arise from design for assembly and disassembly are consumer factor related. For example, being able to easily disassemble a product is not sought after when assembling for reliability, quality, and safety [5–10]. These factors are very important and have to be considered.

When companies see the high startup costs associated with design for disassembly programs, they become discouraged. If a part costs too much to disassemble, so that the cost to disassemble exceeds the actual worth of the part, it is not likely the company will disassemble the product. In the United States, there is no current legislation that mandates companies to disassemble the product, and until legislation occurs, companies in the United States tend to shift away from DFD and the recycling of their products [5–7]. It may cost money to start designing for disassembly, but in the long run it may offer big incentives for companies. An example of this is given by Digital Equipment Corporation.

Digital Equipment Corporation recycles thermoplastic resin from the housings of old computers that customers have traded in. It also recycles cathode ray tubes from used computers. This caused Digital to save more than $1 million in hazardous waste landfill fees. Designing for disassembly helped Digital and the environment.

If companies are going to implement DFD to cut landfill contributions, products must be designed with the environment in mind right from the start. Materials need to be selected and combined in certain ways so recycling and disassembly are made easier. Initial selection of materials is one of the most important facets of DFD. A good example of material selection is the two basic types of plastics used in cars, which are thermoplastic and thermoset. The former is used commonly in dashboards, consoles, and other interior parts. Thermoplastic, once the automobile is scrapped, can be melted down like wax and remolded into something else. Each day there is more and more experimentation with thermoplastics since they are so easy to recycle. Automobile companies are looking for more ways to use thermoplastics in substitution for thermosets. At present there are many types of thermoplastics, which include polystyrene, polyvinyl chloride, and polypropylene (see Chapters 5 and 6). A system has to be developed where all these materials must be separated from each other because an impurity can weaken a part. One system being adopted by many companies is an identification scheme. This identification scheme was suggested by the Society of Automotive Engineers. This scheme involves codes that are carved into the plastic molds. The actual part will be molded with an identification code. This will make recycling easier, since parts can be identified quickly and accurately.

Another consideration for DFD is the actual assembly itself. Fasteners have to be considered wisely. Parts should not be welded or glued together, because this makes them harder to separate from each other and from the actual adhesive compound. A solution to this problem is snap-fit type fasteners or screws that can be disassembled quickly. Screws can be removed quickly, but if the part has a lot of screws, the time to disassemble the part will be high. Snap-type fits make it easier and quicker to disassemble a part than screws. The problem with fasteners is that the product should not be built and designed for the customers to disassemble it themselves. It should be left for the company to disassemble the part. The topic of fasteners leads into the next principle for design for disassembly, the reduction of parts in a product.

If a product has a vast number of parts, it takes longer for the product to be disassembled. The fewer the parts, the faster the disassembly. Subassemblies have to be combined in some way so the number of parts can be consolidated into a smaller number of parts. This leads to other benefits such as savings in energy costs of production since there are fewer pieces to manufacture. A good example of this is the Trilliant circuit breaker box. The original circuit box design was made out of metal and consisted of 173 parts. The new design used injection-molded plastic and has reduced the number of parts to 42 [11].

At present, there are not that many products on the market that are designed for disassembly. Therefore, this chapter presents some of the methodologies that can help industry to step forward and take its share in the era of green technology. First a detailed discussion of the design process is provided. Problems inherent in the design process are identified, followed by a new technical approach to solving these problems. The problems are stated with emphasis on the bottlenecks of decision making, and the solutions are detailed in a step-by-step procedure.

7.4   THE DESIGN PROBLEM

The level of complexity associated with defining a "single point design" including its analytical validation is low compared to the task of qualifying a "very good" or potentially "best" design. The definition of "best" is associated retrospectively with competitive success and in the early stage is often elusive [12,13]. The achievement of "best" is likely to require the competing decisions of many groups, that is, marketing strategy, production, legal, environmental, upper management, etc., each with different sets of priorities that may not easily normalize to simple metrics of optimum, such as cost. The design process from this perspective becomes a subset of business planning, and high-quality, accurate, and timely information about product engineering should be considered an essential resource to productive operations.

Decision making is important at the organizational, team, and individual level. It is the fundamental and time-consuming task of product design for many reasons, including:

- The scope of potentially "best" design options expands exponentially as one approaches the level where accurate cost and performance evaluation can begin.
- The product requirements in the early stages may be poorly understood, leading to work on the basis of vague assumptions.
- Errors in judgment at early stages go unnoticed because they are never explicitly recorded for later review.
- Large resources are committed to subtasks, that is, complex finite-element analyses for the purpose of making informed decisions, and there is strong pressure to initiate these subtasks early to demonstrate that work is being done.

The latest high powered desktop simulation tools seem to have escalated these problems. The ability to analyze structures, fluids, magnetics, mechanisms, etc., has made it quite easy to generate extraordinary amounts of analytical results. Although these new methods are powerful and useful, their overuse can divert attention from more fundamental issues of the design process. Some of these are:

- What is a viable concept and how can we converge on a physical representation that best satisfies the functional requirements?
- What minimum set of information (i.e., far short of a complete design) could be used as a basis to compute accurate estimates of processing costs?
- What methods and information are available at the concept level that could help us rank our list of design options?
- At what point can intensive analysis and optimization be used to best advantage?
- Explain in summary what chain of reasoning and assumptions have led to the current state of the design process and why a particular design feature was chosen.

In theory the design problem is open-ended. It is unrealistic to lay siege to all possible options because of resource limits, yet there is often pressure to justify that type of approach. There are few simple ways to tell when a "best" solution has been found, or conversely, if a marginal design improvement will offset the required effort.

In practice the process should end when it can be demonstrated with "high confidence and reasoning" that a sufficiently broad scope has been considered and that the probability of existence of a better design solution is low. Significant improvements would be realized with next-generation CAD systems, which include information representing the design intent, and this is best served by a structured definition of needs. It is therefore the objective of this chapter to describe a new methodology to achieve this goal. This methodology has a fundamental basis in recent theoretical developments in the science of design and would be implemented best by an interactive graphics-based software application. This application would assist in the definition of design parameters as an explicit means to satisfy a progressively decomposed set of requirements and would provide detailed, concept-level associations of requirements to available options. It would make intensive use of existing expert systems, axiomatic theory of design, database formats, and analysis tools, and would provide new symbolic representations to clarify the decision structure inherent in the design process as discussed next.

7.5   TECHNICAL APPROACH

7.5.1   Axiomatic Design

The axiomatic approach to design eliminates the need for an exhaustive search of all possibilities by focusing on principles available for decision making in design qualitatively [14–17]. The goal of axiomatic design (AD) is to establish a scientific foundation for the design field so as to provide a fundamental basis for the creation of products and processes. This goal is achieved by defining a set of functional requirements that satisfy the perceived needs for product and then creating a tangible entity in the physical domain that satisfies the stated functional requirements. The AD approach establishes decision rules and their derivatives so that they can be interpreted and applied by computers. This systematic approach of making rational decisions based on the synthesis of various pieces of information is very helpful in the abstract phase of a design. This approach is based on two general principles as follows [14]:

1. The independence axiom. Seek to minimize the degree of dependence of functional requirements (FR), and rank each design by preference in the following order: uncoupled, decoupled, and coupled.

2. The information axiom. Seek to minimize the amount of information needed to define a complete design and rank each design by preference of lower information content.

The three cases of relative independence are represented in matrix format in Figures 7.1 to 7.4. This matrix representation is used to identify dependencies, where any given functionality is said to depend on a specific subset of design parameters. Figure 7.1 represents an example of the uncoupled case—the simplest, where each function is dependent on a single design parameter (DP). This situation occurs rarely and amounts to the trivial selection of design parameters in any arbitrary order.

The second case is the most general circumstance and is the primary focus of this chapter. Organizing functional requirements in increasing order of dependencies results in a priority of tasks that is most efficient for the overall design solution. Design parameters should therefore be sized to meet the "earlier" functional requirements, which are those at the top of the matrix shown in Figure 7.2. As design parameters become more progressively defined, the remaining requirements can be satisfied by sizing a minimized remaining set of variable design parameters.

Figure 7.1 Uncoupled designs.

	DP1	DP2	DP3	DP4
FR1	X			
FR2		X		
FR3			X	
FR4				X

Figure 7.2 Decoupled designs.

	DP1	DP2	DP3	DP4
FR1	X			
FR2	X	X		
FR3	X	X	X	
FR4	X	X	X	X

A simple example of this is shown in Figure 7.4. The functional requirements are engine power and mechanical attachment. The design parameters are piston size and mounting features. The preferred order is to first calculate those parameters that define the power output, that is, piston diameter, stroke, eventual block size, etc. Following the definition of these parameters the attachment arrangement (i.e., mounting bolts) can be determined from the overall size and weight previously determined. This order was selected since the mounting feature does not affect the power output of the motor (it can be ignored during the design of power components) but does depend on the overall motor size.

This preferred order of functional requirements may not be as apparent in the concept stage, where the very inclusion of given design parameter sets is dependent on design concepts yet to be defined, and therefore unknown. As concepts become more progressively defined during the design process, the degree of coupling between functional requirements and design parameters is one way of evaluating the complexity of a concept, and a useful metric in ranking the list of possible design concepts. Identifying the preferred order of decoupled functional requirements for each concept provides a basis for organizing the growing task of trade study comparisons.

The third case is that of highly coupled designs and is represented by Figure 7.3. Design problems in this category generally require specialized forms of optimization to solve. Whenever possible it is desirable to avoid the expense and complexity of such intensive efforts by rearranging or redefining the requirements to result in a decoupled design as shown in Figure 7.2.

The second axiom roughly translates to simplicity. A unit of information implies an associated feature of dimension, tolerance, material processing, assembly steps, etc. One can generally assume that each feature incurs an incremental cost in terms of process setup, machine time, inspection, and possible rejection. This relationship between information and cost is obviously not linear. All features will not generally incur the same cost. Nonetheless, it can provide evaluation estimates at an early stage when cost computations are not practical, and can be extended to become the basis for cost estimation with further refinement. It also provides feedback to the designer on the relative simplicity of various design options.

Figure 7.3 Highly coupled designs.

	DP1	DP2	DP3	DP4
FR1	X	X	X	X
FR2	X	X	X	X
FR3	X	X	X	X
FR4	X	X	X	X

Figure 7.4 Decoupled design example.

	Piston Diameter	Mounting Features
Power	X	
Mech. Attach	X	X

7.5.2   Functional Requirements

A real-world product requirement is used as the starting point for the example in this discussion [18]. This requirement comes from a request for proposal (RFP) from the U.S. Navy for a lightweight, quiet, portable generator with a capacity of 0.5 kilowatts. To begin, a formal statement of need or functional requirement (FR) set is identified accompanied by a claim that "*all design activity should be requirements driven.*" The FR set for this example is shown in Table 7.1.

An FR implies a need in the imperative, such as "Provide 0.5 kW of power" or "Provide 110 volts." Functional requirements are said to exist in the abstract functional domain and generally have a multitude of possible solutions that exist in the physical domain. FRs take a number of forms that can be further distinguished. The primary FR set defines the initial need for product function, and a subset of this termed the root FR set forms the core of this definition. These distinctions are somewhat arbitrary but will become apparent in later discussion.

At this stage an FR is usually quite abstract, due to the diverse wants of human nature. It is therefore useful to decompose an FR set into a set of less abstract requirements, for each of which a solution can be more readily proposed. These lower level requirements are themselves "requirements of function," and where they refer to a physical component or design concept they are termed a design requirement (DR). DRs are specific to various possible concepts and are said to be dependent or conditional. That is, there exists at least one other design solution that satisfies the primary FR set and does not include that particular design feature. The relevance of a DR thus depends on its associated feature(s) being included in the final design.

A functional requirement is said to be resolved if its decomposition set (lower level) functional requirements are met or resolved. This relationship could take the form of a Boolean predicate that represents an FR by its decomposition set. Note

Table 7.1. Functional Requirements for Portable Generator

FR#	Requirements
1	Portable electric generator
2	0.50 kW Power
3	110 volts
4	Direct current
5	Alternating current, 60 HZ
6	Uses fuel currently available
7	Quiet operation (40 db @ 10 meters)
8	Reliable operation (8,000 hr continuous)
9	Designed for simple automated assembly
10	Harsh environment: dust, humidity, mishandling (-40 °C to +60 °C, 40 g shock)

that the decomposition process fulfills an important need in itself, to direct the focus of design effort to where it is most relevant. This needs-driven approach might also be termed a "pull" system, where efforts are directed by requirements to the physical domain on the basis of need.

A practical example of an electric motor demonstrating the applicability of these concepts is presented in the following section.

7.6   THE ELECTRIC MOTOR

To support DFA concepts, knowledge-based expert systems (KBES) that coordinate different users toward designing for manufacturability are needed. These programs contain information on each component of the product under consideration for both design and manufacturing. Such component information and the assembly operation information are stored in the knowledge base. The methodology will be applied to the design and manufacturing of an electrical motor consisting of 18 components [19]. For automatic assembly, contributions from the design and manufacturing groups can generate improvements in the following areas [20].

1. Simplicity: Reduce number and variety of parts.
- Combine several functions in one part.
- Employ identical parts.
- Form assembly families (or subassemblies).
2. Improvement: Provide alternatives for assembly planning.
- Enable random assembly sequences.
- Enable subassembly sequences if cannot be avoided.
- Avoid compulsory sequences.
3. Efficiency: Facilitate assembly execution.
- Design for insertion suitability.
- Design for joining suitability.
- Avoid nonassembly work.

Knowledge-based expert systems are used in assisting engineers in designing better products and processes. This assistance can range from the conceptual design stage to the final inspection stage. A knowledge-based expert system consists of three major modules: (1) a knowledge base, (2) an inference engine, and (3) a database.

The knowledge base (KB) contains information on how to improve products by studying the characteristics of each part that makes up the product and the required production processes. The inference engine is a general-purpose computer program that uses KB and database and links all rules and facts of the product and process effectively to infer to certain solution. It evaluates the solutions and provides recommendations to the user.

The database stores component characteristics information. It is open for updated design modifications from different users and has updating mechanisms for new data. In addition, possible alternative component designs can be included in the database by specifying its functional, dimensional, and material parameters [21].

Now, consider a hypothetical design of the electric motor as shown in Figure 7.5 and establish a statement of need or functional requirement set. All design activity is claimed to be requirements driven. The functional requirement (FR) and the design parameter (DP) sets for the assembly process of this motor are:

1. FRs for Assembly

FR1 = fix rear housing
FR2 = assemble rear bearing
FR3 = assemble shaft
FR4 = assemble front bearing
FR5 = assemble front housing
FR6 = lock front housing with rear front housing
FR7 = assemble pulley-fan-shoulder
FR8 = assemble lock washer

2. DPs for Assembly

DP1 = place rear housing in the fixture with its opening facing up
DP2 = insert rear bearing into rear housing
DP3 = insert modified shaft into rear bearing
DP4 = slide front bearing over shaft
DP5 = insert front housing over the shaft and rear housing
DP6 = insert and tighten two through-bolts over front housing and rear housing
DP7 = slide pulley-fan-shoulder (modified) over shaft
DP8 = slide lock washer over shaft

The design process is mimicked where the particular rules that make up the system originate from AD. These concepts are programmed as modules of the knowledge-based expert system as shown in Figure 7.6 and described next.

Evaluation. This module takes different designs and evaluates the optimum design. It consists of largely five sections as follows:

Section 1. Contains description of the program and definitions of predicates.
Section 2. Contains axiom 1 and axiom 2 (the independence axiom and information axiom described earlier). The two axioms were used as bases in generating KBES rules that support DFA. Especially the first axiom (design functional requirements independence) was incorporated at every hierarchical level of design parameters (components that satisfied assembly process constraints were chosen during product design). If component design violates the first axiom, the outcome can be coupled design during the manufacturing stage because manufacturing steps considered during the later stages of production can affect the work done at earlier stages.

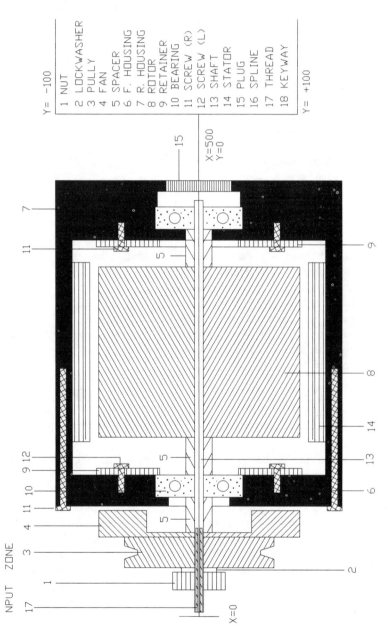

Figure 7.5 Hypothetical electric motor.

INPUT ZONE

Y= -100

1 NUT
2 LOCKWASHER
3 PULLY
4 FAN
5 SPACER
6 F. HOUSING
7 R. HOUSING
8 ROTOR
9 RETAINER
10 BEARING
11 SCREW (R)
12 SCREW (L)
13 SHAFT
14 STATOR
15 PLUG
16 SPLINE
17 THREAD
18 KEYWAY

Y= +100

X=500
Y=0

X=0

142

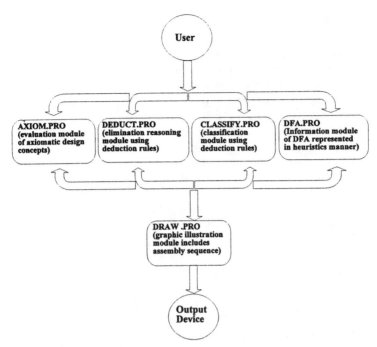

Figure 7.6. Structure of the expert system.

Section 3. Contains feasibility check mechanism and independence measure mechanism. A design is feasible if all of its design parameters fall within the required functional requirement range. The independence of a design is assigned here depending on the existence of nonzero entity of the design matrix.

Section 4. Contains information measure mechanism. It calculates various kinds of information associated with different designs. It is capable of obtaining information measure on surface finish, geometry, and machining time, and finds total information associated with each design.

Section 5. Contains the functional requirements and design parameters (described earlier).

Classification. When a designer has much information on many different subjects, but cannot make a decision due to the vast amount of data, a systematic way of decision making is needed. This module figures out which attribute from each domain is best suited by asking queries, then making deductions from the replies given. Each domain is classified in different categories (e.g., material), which are then broken up into subcategories (e.g., babbitt_alloy). Each domain is described by a number of attributes in the form of rules and conditions that are stored in the following format:

rule(ID#,"category", "sub-category", [satisfied conditions])
condition(ID#, "attributes")

As an example, consider the following rules and conditions for a shaft material:

rule(1, "material", "babbitt_alloy", [1,2,3,4,5])
rule(2, "material", "copper_alloy", [6,7])
condition(1,"has_excellent_embeddability")
condition(2,"has_excellent_compatibility")
condition(3,"has_poor_load_capacity")
condition(4,"has_poor_strength_at_high_temperature")
condition(5,"has_poor_fatigue_strength")
condition(6,"has_good_load_capacity")
condition(7,"has_good_score_resistance")

Because the sequence of selecting and applying rules is not specified, the inference mechanism determines which rules are applicable and invokes corresponding actions. Information such as the following can be obtained:

- Various types of fastening methods
- Components material selection
- Surface finish due to different machining process
- Surface hardness due to different heat treatment
- Load capacity of different bearing type

Reasoning. This module is designed to make deductive reasoning on a large knowledge base in order to arrive at certain conclusions. The knowledge base contains DFA techniques on locking fasteners, bearing mounting methods, fan design, housing design, spacer design, and material selection. In the knowledge base, a series of IF-THEN rules are coded. These statements simulate an expert's "if-certain situation, then-take certain action" decisions. The objective of this inference mechanism is to match given conditions with AD and DFA considerations to yield a better design. Each fact of the rule base is identified with a time stamp to indicate the time when it was inferenced. During matching rules, this time stamp is used to select which rule to execute. All rules are stored in the clause section in the following format:

IF(L):- Condition$_i$ (T$_i$)
L = [ID#, T$_i$]
THEN(L), Conclusion$_j$

Each member of T_i is the time stamp associated with certain **Condition$_i$** and **ID#** is the identifying number of rules. The time stamp T_i is used to keep track of the members of the list **L** after each iteration of the algorithm. It is updated after the list **L** is checked for the condition.

Guidance. This module is capable of guiding the user to design a product based on DFA concepts. Its database consists of guidelines for better product design and corresponding assembly process. By examining all branches of the data base, the user can obtain useful information on better design of each component that satisfies assembly constraints. User-friendliness is incorporated well so that through a simple interface, the following database (which is represented in heuristic manner) may be obtained:

- General assembly-concerned design
- Orientation motion-concerned design
- Transportation motion-concerned design
- Merging motion-concerned design
- Joining motion-concerned design and components information

By using this module, not only is design improvement of each component possible, but also combining two or more components into one is considered. Examples of using the guidelines would be the decoupling of coupled designs, minimization of functional requirements, integration of physical parts, standardization, symmetry, largest tolerance, etc. For example, the integration of physical parts might include elimination of screws, combination of parts with protrusions, etc.

Graphical Presentation. This module uses Borland Graphics Interface (BGI) mode for graphics illustration of user's design [22]. It is designed to be iterative during backward chaining so that drawing commands can be repeated as many times as the user wishes. All components that make up a product are represented by texture-filled polygons and are identified numerically. These components are called and placed within the X and Y axis limits of the screen. Their dimensions, coordinate points, and number of occurrences are chosen by the user. The assembly sequence is included in the program by directing the user to the next component after creating a component.

Application of the Expert System. To explore the capabilities and applicability of KBES, consider the motor shown in Figure 7.5. The figure illustrates how a nut and lock washer are used to lock pulley and fan. Even though this is a widely used method, it is a poor design for automatic assembly consideration because of the awkward assembly motions required. This method also requires a plug at the rear housing that allows holding the shaft during tightening the nut. The plug in the rear housing can be designed to be snap fitted or thread fitted to the rear housing.

External threads on the shaft begin at the left end of the shaft. During the process of inserting a component (spacer, bearing, front housing, fan, and pulley) over the shaft, damage to threads can occur. Even though the pulley and fan are press fitted over the shaft, they can slip around the shaft during operation. In order to develop a different locking fastener that improves design efficiency, REASONING was run. Current design conditions were given through the interface of the data file. These conditions were as follows: The lockwasher is held in place in the final assembly by being slid over the shaft and sandwiched between the pulley and its retaining nut. Since, during assembly, the washer is a loose part, REASONING recommended incorporating it into the nut for a single assembly, a nut/lock washer combination as shown in the improved design of Figure 7.7. The CLASSIFICATION's fastener

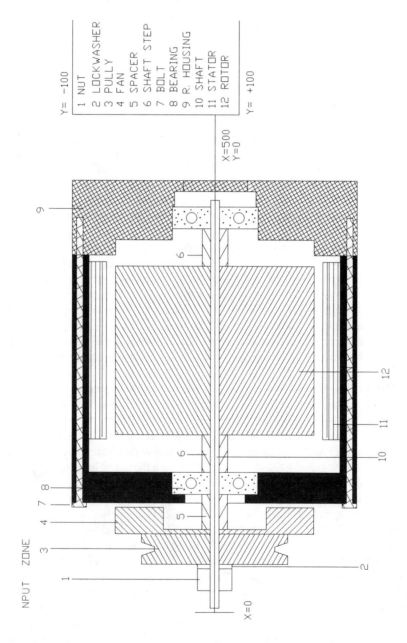

Figure 7.7 Improved design of the electric motor.

domain responded to this recommendation by pointing out a number of problems including required machining, assembly motion, locking force, and fixturing subassembly. Although solutions to these problems are not yet incorporated in the expert system, they do alert the designer to their presence and their need for immediate solutions.

Discussion of Results.   To examine the efficiency of KBES, the hypothetical motor is analyzed at various subassembly levels and with substantial improvements. Both the hypothetical and improved motors were compared using Boothroyd's DFA techniques. This involves studying two areas: time spent for handling each component and time spent to assemble the component. The hypothetical motor was divided into its basic components listed in Table 7.2.

The table also shows the operation time for each component (T_o) and the number of components that cannot be avoided or eliminated for further improvement of the motor (C_a). Using Boothroyd's technique, the design efficiency was computed as 12%.

Many components have been changed in this design to facilitate easier assembly. The rear housing was made shorter in length for easier insertion of rear bearing and shaft. The shaft now has bigger diameter, which replaces the function of two spacers. The front housing is made longer to provide total housing length due to decrease in rear housing length. There are only two through-bolts to tighten the two housings, and this matches the minimum number of through-bolts that cannot be

Table 7.2. Hypothetical Motor DFA Analysis

COMPONENT	T_o	C_a	COMPONENT	T_o	C_a
(1)rear housing	1.5	1	(10)front retainer	7.1	0
(2)rear bearing	2.6	1	(11)front retainer short screws	41	0
(3)rear retainer	7.1	0	(12)turn f_housing over & insert to r_housing	9.5	-
(4)rear retainer short screws	41.	0	(13)through bolts	41.	2
(5)spacer3-shaft-rotor	12.	0	(14)spacer1	12.	0
(6)spacer2-shaft-rotor	12.	0	(15)fan	4.	1
(7)shaft-rotor-spacer2	3.8	1	(16)pulley	4.6	1
(8)front housing	1.5	1	(17)lock washer	2.6	0
(9)front bearing	2.6	1	(18)nut	13	0
			(19)plug	3.6	0

Note. ΣT_o = 220.5 seconds, ΣC_a = 9 and hence design efficiency = $(3\ \Sigma C_a)/\Sigma\ T_o$ = 0.12 ==> 12 %.

Table 7.3. Modified Motor DFA Analysis

COMPONENT	T_o	C_a	COMPONENT	T_o	C_a
(1)rear housing (m)	1.5	1	(6)through bolts (m)	27.0	2
(2)rear bearing	2.63	1	(7)pulley-fan (m)	7.25	1
(3)shaft-rotor (m)	3.75	1	(8)lock washer	2.63	0
(4)front bearing	2.63	1	(9)spline	6.13	0
(5)front housing (m)	8.0	1			

Note. $\Sigma T_o = 61.52$ seconds, $C_a = 8$ and hence, design efficiency $= (3\Sigma\ C_a) / \Sigma\ T_o = 0.39 \Longrightarrow 39\ \%$

avoided. The biggest change is in the pulley, fan, and first spacer, where all were integrated as one component. The available molding technology can facilitate this integration process and hence can save substantial assembly time. The term "mold" here refers to its being manufactured as one piece. This is the design goal—that is, reducing the number of parts and assembly operations. However, the pulley may be molded out of an "engineering" plastic such as Delrin, or it could be made as a zinc die casting or similar technique. The motor torque specification and projected quantities would dictate the material and manufacturing method. To analyze the improved design of Figure 7.7, consider the list of components shown in Table 7.3.

By implementing KBES, the design efficiency was improved from 12% (hypothetical design) to 39% (modified design). Total operation time decreased significantly from 220.5 seconds to 61.5 seconds. The cost due to modifications in existing component designs can be compensated by this substantial decrease in assembly.

Examining the improved design of Figure 7.7 and Table 7.3 shows that the design can still be further improved. The expert system recommended three improvements, and they are shown in the revised design of Figure 7.8. These improvements include:

1. Modifying the spacer so that the rotor is locked to the front housing by the same nut that holds on the fan and pulley assembly.
2. Reverse the two through-bolts to make them shorter, hence reducing manufacturing costs.
3. Eliminate the spline (not shown in drawing) in the shaft, which is expensive to machine.

The change of bolts' length and position promoted the idea of completely eliminating the bolts; although this recommendation was not made by the expert system, it should be investigated. It is proposed to update the production rules of the expert system to reflect this change. An encouraging factor is the increase in the design efficiency as obtained from Table 7.4 and the following calculations:

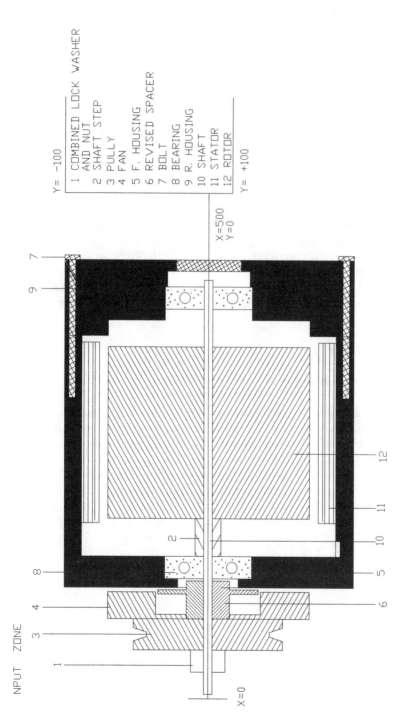

NPUT ZONE

Y= -100

1 COMBINED LOCK WASHER
  AND NUT
2 SHAFT STEP
3 PULLY
4 FAN
5 F. HOUSING
6 REVISED SPACER
7 BOLT
8 BEARING
9 R. HOUSING
10 SHAFT
11 STATOR
12 ROTOR

Y= +100

X=500
Y=0

X=0

Figure 7.8 Revised design of the electric motor.

149

Table 7.4. Redesigned Motor DFA Analysis

COMPONENT	T_o	C_a	COMPONENT	T_o	C_a
(1)rear housing (m)	1.5	1	(5)front housing (m)	7.0	1
(2)rear bearing	2.63	1	(6)pulley-fan (m)	7.25	1
(3)shaft-rotor (m)	3.75	1	(7)lock washer	2.63	0
(4)front bearing	2.63	1	(8)spline	6.13	0

Note. $\Sigma T_o = 33.5$ seconds, $\Sigma C_a = 6$ and hence design efficiency = $(3 \ \Sigma C_a) / \Sigma T_o = 0.54 ==> 54\%$.

As can be seen, the design efficiency has improved significantly from 12% (theoretical) to 54% (revised). Total operation time decreased significantly from 220.5 seconds to 33.5 seconds. As an example, operation time of housing has decreased to 7 seconds because of the symmetry around the assembly axis. The cost due to the redesign of existing components such as the front and rear housings can be compensated by this substantial decrease in assembly. It is very important to note that the material of the housings, such as cast iron, may not allow for use of such features as snap-fit. Therefore, the inclusion of the type of the material into the expert system along with its properties and constraints would dictate the design and manufacturing method.

7.7   EXAMPLES

7.7.1   Electric Hair Dryer

An electric hair dryer may involve at least 27 assembly operations, as illustrated by the number of parts shown in Figure 7.9. These may be unnecessary since operations and parts can be combined and eliminated. For example, it will be very difficult to assemble and disassemble the element bracket while the motor and fan is in place. It is more practical to break the hair dryer down into a small number of subassemblies and to carry out individual studies of each. The order of subassembly may not be a very important factor. In this case, the operations can be divided into subassemblies to ease the assembly and disassembly of the product. Each subassembly consists of various parts that will be assembled in the frame as follows:

1. Frame subassembly. This contains the main frame that includes all the subassemblies.
2. Element bracket subassembly.
      a. Heating element coil
      b. Capacitor
      c. Heating sensor
3. Motor subassembly
      a. Motor
      b. Motor flange
      c. Fan
      d. Motor wire

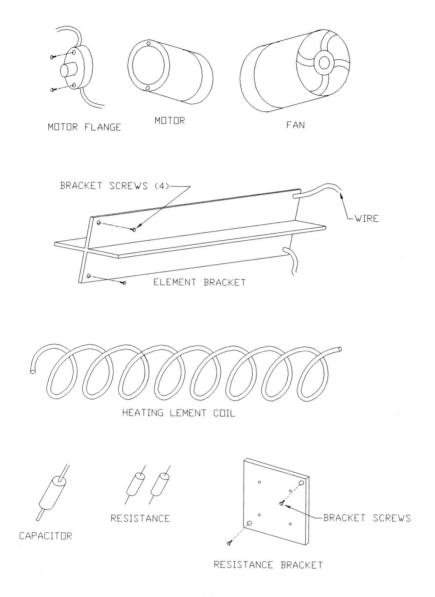

Figure 7.9a Some parts of the original design of the hair dryer.

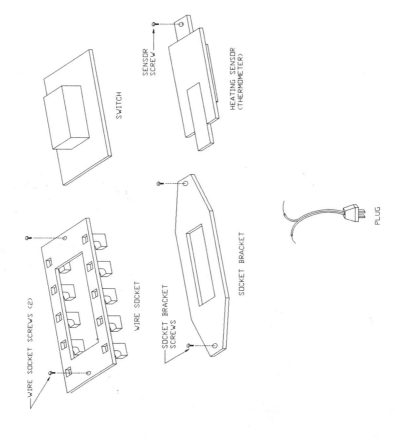

Figure 7.9b Additional parts of the original design of the hair dryer.

4. Resistance bracket subassembly
    a. Resistance and bracket
    b. Wire
5. Switch subassembly
    a. Switch
    b. Wire socket
    c. Socket bracket
    d. Wire

Procedure for Assembly.

1. Frame subassembly
2. Element bracket subassembly
3. Motor subassembly
4. Resistance bracket subassembly
5. Switch subassembly
6. All wire connection
7. Plug connection
8. Frame cover
9. Frame cover screws
10. Front side cover
11. Front side cover screws
12. Back side cover
13. Back side cover screws

Procedure for Disassembly of the Revised Design

1. Air nozzle
2. Front side cover (redo snap fit)
3. Back side cover (redo snap fit)
4. Frame cover
5. Switch subassembly
6. Switch components
7. Motor subassembly
8. Motor parts
9. Element bracket subassembly
10. Element bracket components

Procedures That Can Be Implemented by Industry for DFD. The most obvious way in which the disassembly process can be facilitated at the design stage is reducing the number of parts to a minimum. Also, before a design can be analyzed, it must be decided whether the disassembly process is carried out manually or with the aid of automation. Another factor to be considered in DFD is the assembly of the parts since easy assembly will result in easy disassembly. The designer should provide a layer design fashion, where each part can be placed on top of the previously assembled one.

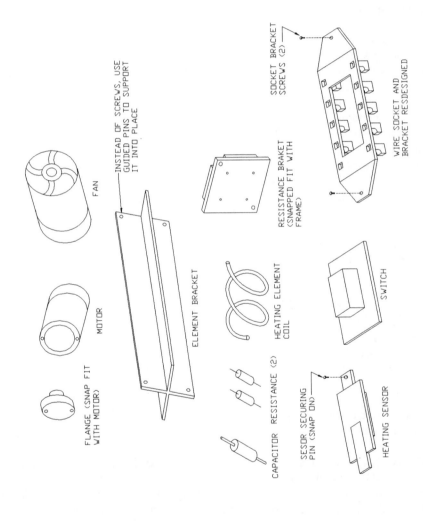

FAN

MOTOR

FLANGE (SNAP FIT
WITH MOTOR)

INSTEAD OF SCREWS, USE
GUIDED PINS TO SUPPORT
IT INTO PLACE

ELEMENT BRACKET

RESISTANCE BRAKET
(SNAPPED FIT WITH
FRAME)

HEATING ELEMENT
COIL

CAPACITOR RESISTANCE (2)

SWITCH

SESOR SECURING
PIN (SNAP ON)

HEATING SENSOR

SOCKET BRACKET
SCREWS (2)

WIRE SOCKET AND
BRACKET RESDESIGNED

Figure 7.10a Redesigned main frame and cover of the hair dryer.

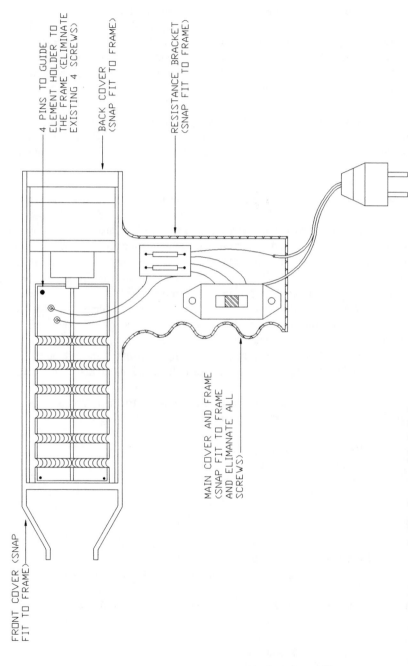

FRONT COVER (SNAP
FIT TO FRAME)

4 PINS TO GUIDE
ELEMENT HOLDER TO
THE FRAME (ELIMINATE
EXISTING 4 SCREWS)

BACK COVER
(SNAP FIT TO FRAME)

RESISTANCE BRACKET
(SNAP FIT TO FRAME)

MAIN COVER AND FRAME
(SNAP FIT TO FRAME
AND ELIMANATE ALL
SCREWS)

Figure 7.10b Some redesigned parts of the hair dryer.

The DFA/DFD analysis shows that assembly procedures in the revised design are much easier than the original design because of the environmentally integrated guidelines described earlier. By redesigning the hair dryer, the design is kept extremely simple. This is implemented by using as few materials as possible and by incorporating many functions into one function without affecting the design intent. By providing snap-fits, a 13-screw operation was eliminated. In using snap-fits, tight tolerance design principles need to be applied to reduce the use of fasteners and to keep the separation process simple.

Original Design Efficiency of the Electric Hair Dryer. By conducting a similar analysis to the electric motor described in Section 7.6, the following information is used to compute the design efficiency: $\Sigma T_o = 416.38$ seconds, $\Sigma C_a = 37$, and hence the overall design efficiency = $(3 \Sigma C_a)/\Sigma T_o = 0.27$ (27%).

Redesigning the Electric Hair Dryer. The most obvious way in which the disassembly process can be facilitated at the design stage is reducing the number of parts to a minimum. Also, before a design can be analyzed, it must be decided whether the disassembly process is carried out manually or automatically. Another factor to be considered in DFD is the assembly of the parts, because easy assembly will result in easy disassembly. The designer should provide a layer of fashion design, where each part can be placed on top of the previously assembled one.

By redesigning the hair dryer, the design is kept extremely simple. This is implemented by using as few materials as possible and by incorporating many functions into one function without effecting the design intent. By providing snap-fits, a 13-screw operation was eliminated. By using snap-fits, tight tolerance design principles need to be applied to reduce the use of fasteners and to keep the separation process simple.

Some redesigned components of the new electric hair dryer are shown in Figure 7.10. The improvements were achieved in the following areas:

- Provide a snap-fit design in:
  a. Resistance bracket
  b. Frame cover
  c. Front side cover
  d. Back side cover
- Eliminate 13 screws.
- Eliminate two reorientations.
- Revise wire soldering by using crimp lugs.
- Combine the wire socket and the socket bracket to form one piece.

and the procedure of disassembly of the revised design is as follows:

1. Air nozzle
2. Front side cover (Redo snap-fit)
3. Back side cover (Redo snap-fit)
4. Frame cover

5. Switch subassembly
6. Switch components
7. Motor subassembly
8. Motor parts
9. Element bracket subassembly
10. Element bracket components

In conclusion, the revised design reduced the total time from 416.38 seconds to 158.84 seconds and reduced the number of parts to 34, leading to an improved design efficiency of 64%.

7.7.2   Electric Hair Clipper

Consider the electric hair clipper shown in Figure 7.11. This clipper requires 26 assembly operations as illustrated by the number of parts shown in Figure 7.12. Certainly, operations and parts count can be reduced if DFA/DFD guidelines are properly implemented. For example, the operations can be divided into subassemblies as follows:

1. Drive plate subassembly
      a. Drive plate
      b. Conductor plate
      c. Spring holder
2. Switch subassembly
      a. Switch
      b. Switch socket
      c. Electric wire
3. Bracket subassembly
      a. Core / coil bracket
      b. Socket bracket

Procedure for Assembly.

1. Frame subassembly
2. Drive plate subassembly
3. Springs
4. Alignment screw
5. Switch subassembly
6. Core/coil assembly
7. Electric wire
8. Plug connection
9. Core bracket pins
10. Bracket subassembly
11. Switch knob (on/off)
12. Frame cover

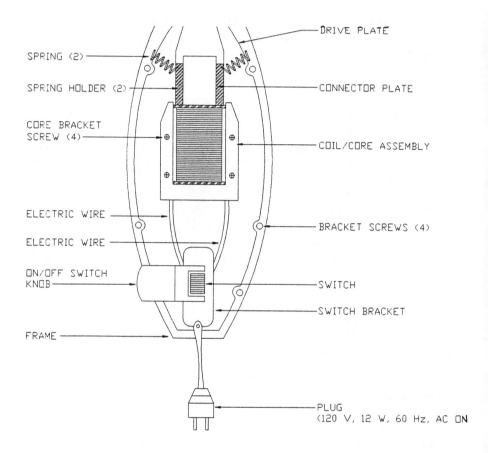

Figure 7.11   Original design of the electric hair clipper.

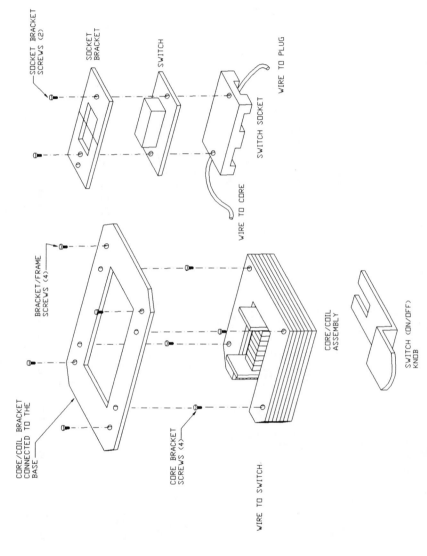

Figure 7.12a Some parts of the original design of the hair clipper.

SOCKET BRACKET SCREWS (2)

SOCKET BRACKET

SWITCH

WIRE TO PLUG

SWITCH SOCKET

WIRE TO CORE

BRACKET/FRAME SCREWS (4)

CORE/COIL ASSEMBLY

SWITCH (ON/OFF) KNOB

CORE/COIL BRACKET CONNECTED TO THE BASE

CORE BRACKET SCREWS (4)

WIRE TO SWITCH.

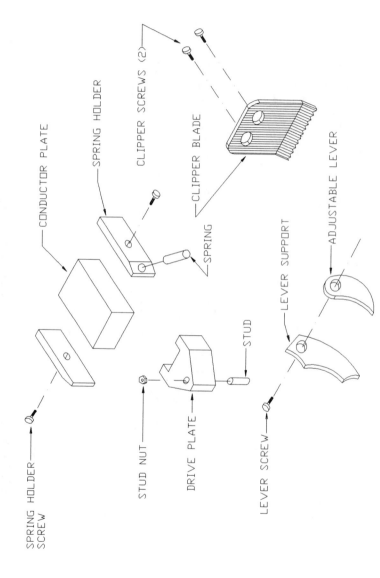

Figure 7.12b Additional parts of the original design of the hair clipper.

13. Clipper blade
14. Clipper blade screw
15. Lever support
16. Adjustable lever
17. Lever screw

Procedure for Disassembly of the Revised Design.

1. Clipper blade and lever
2. Frame cover
3. Switch knob
4. Bracket subassembly
5. Core/coil assembly
6. Switch subassembly
7. Springs
8. Drive plate subassembly

Original Design Efficiency of the Electric Hair Clipper. By conducting analysis similar to that of the electric motor described in Section 7.6, the following information is used to compute the design efficiency: ΣT_o = 345.81 seconds, ΣC_a = 26 and hence the overall design efficiency = $(3\ \Sigma C_a)/\Sigma\ T_o$ = 0.23 (23%).

Redesigning the Electric Hair Clipper. Some redesigned components of the new electric hair clipper are shown in Figure 7.13. The improvements were achieved in the following areas:

-     Provide a snap-fit design on:
  a. Frame cover
  b. Socket bracket
  c. Coil/core bracket
-     Eliminate 16 screws
-     Eliminate 2 reorientations
-     Revise most wire soldering by using crimp lugs

and the procedure of disassembly of the revised design is as follows:

1. Clipper blade and lever
2. Frame cover
3. Switch knob
4. Bracket subassembly
5. Core/coil assembly
6. Switch subassembly
7. Springs
8. Drive plate assembly

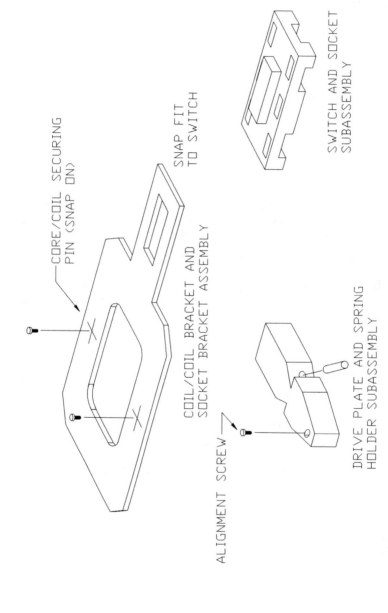

CORE/COIL SECURING
PIN (SNAP ON)

SNAP FIT
TO SWITCH

COIL/COIL BRACKET AND
SOCKET BRACKET ASSEMBLY

SWITCH AND SOCKET
SUBASSEMBLY

ALIGNMENT SCREW

DRIVE PLATE AND SPRING
HOLDER SUBASSEMBLY

Figure 7.13 Some redesigned parts of the hair clipper.

In conclusion, the revised design reduced the total time from 341.81 seconds to 112.20 seconds, reduced the number of parts to 21, and hence improved the overall design efficiency to 56%.

7.8   SUMMARY

Great emphasis has been placed on the need to accelerate the process of product design. A number of key concepts in this chapter are useful to achieve this end:

- Competitive business strategies of the future will depend heavily on effective decision making, which will rely on real-time modeling of product performance and resource demands.
- Productivity improvements could be realized by a structured representation of complex information that associates the abstractions of functional requirements to progressively decomposed sets of tangible design parameters; development of a software application for this purpose is within the scope of current technology.
- Current developments in expert systems and theoretical principles of axiomatic design could be used to implement DFA/DFD techniques.

7.9   BIBLIOGRAPHY

1.   Frosch, R.A., and Gallopoulos, N.E. Strategies for Manufacturing, *Scientific American*, 261, pp. 144–152, September 1989.
2.   Product Recyclability must Be Designed, *Ward's Auto World*, p. 21, December 1994.
3.   Funk, J.L., Design for Assembly of Electrical Products, *ASME Journal of Manufacturing Review*, 2(1), pp. 53–60, 1989.
4.   Boothroyd, G., Dewhurst, P., and Knight, W. *Product Design for Manufacture and Assembly* , Marcel Dekker, New York, 1994.
5.   Lund, T., and Kahler, S. Product Design for Automatic Assembly, 2nd European Conf. on Automated Manufacturing/Automan, Birmingham, UK, 1983.
6.   Baldwin, D.F., Park, C.B., and Suh, N.P. An Axiomatic Approach to the Design of a Continuous Microcellular Polymer Processing System II: Shaping and Cell Growth Control, *Proceedings of the ASME IMECE*, MED-Vol. 1/DE-Vol. 85, pp. 267–277, November 1995.
7.   Ishii, K., Lee, B.H., and Eubanks, C.F. Design for Product Retirement and Modularity Based on Technology Life-Cycle, *Proceedings of the ASME IMECE*, MED-Vol. 2-2/MH-Vol. 3-2, pp.921–933, November 1995.
8.   Yan, X., and Gu, P. Assembly/Disassembly Sequence Planning for Life-Cycle Cost Estimation, *Proceedings of the ASME IMECE*, MED-Vol. 2-2/MH-Vol. 3-2, pp.935–956, November 1995.
9.   Li, W. Zhang, C., Wang, H.P.B., and Awoniyi, S.A. Design for Disassembly

Analysis for Environmentally Conscious Design and Manufacturing, *Proceedings of the ASME IMECE*, MED-Vol. 2-2/MH-Vol. 3-2, pp.969-976, November 1995.

10. Gosch, J. Germany's "Green TV" Signals Trend in Set Design, *Electronics*, 65, p. 29, July 1992.

11. Design for Disassembly: It Is the New Environmental Consideration for Products, *Industry Week*, June 17, 1991.

12. Dixon, J.R., and Duffey, M.R. The Neglect of Engineering Design, *California Management Review*, pp. 9–23, 1990

13. Gupta, A.K., and Wilemon, D.L. Accelerating the Development of Technology-Based New Products, *California Management Review*, pp. 24–44, 1990.

14. Suh, N.P. *The Principles of Design*, Oxford University Press, 1990.

15. Jung, J., and Billatos, S.B. The Applicability of Axiomatic Design In Concurrent Engineering, *ASME National Design Engineering Conference*, McCormick Place, IL, March 8–11, pp. 129–142, 1993.

16. Salustri, F.A., and Venter, R.D. An Axiomatic Theory of Engineering Design Information, *Engineering with Computers*, 8, pp. 197–211, 1992.

17. Kim, S.J., Suh, N.P., and Kim, S.G. Design of Software Systems Based on Axiomatic Design, *Robotics and Computer-Integrated Manufacturing*, 8(4), pp. 243–255, 1991.

18. The Department of Defense, Small Business Innovation Research (SBIR) Program, Navy 21, N93-010, "Miniature Electric Generator," FY 1993.

19. Kroll, E., and Lenz, E. A Knowledge Based Solution to the Design for Assembly Problem, *ASME Journal of Manufacturing Review*, 1(2), pp. 104–108, 1988.

20. Billatos, S.B. Guidelines for Productivity and Manufacturability Strategy, *ASME Journal of Manufacturing Review*, 1(3), pp. 164–167, 1988.

21. Jakiela, M. "Intelligent Suggestive CAD Systems, Research Overview," LMP-MIT, 1989.

22. *PDC Prolog Reference & User's Guides* (version 3.2), Prolog Development Center, 1990.

23. Fenner, R.T. Designing Extruder Screws and Dies with the Aid of Computers, *Conference on New Techniques in Extrusion and Injection Molding*, Manchester, UK, pp. 114–118, June 1974.

24. Fisher, E.G. *Extrusion of Plastics*, Plastics and Rubber Institute, Wiley & Sons, New York, 1976.

25. Lange, K., and Nester, W. "Ceramic Extrusion Dies—Analysis and Application," pp. 118–127, NAMRC, 1985.

26. Riggle, D. Component Recycling for Old Computers, *Biocycle*, 34, pp. 67–69, June 1993.

27. Constance, J. Can Durable Goods Be Designed for Disposability?, *Mechanical Engineering*, 114(6), pp. 60–62, June 1992.

28. Gosch, J. Will the EC Follow Germany's Lead on Computer Recycling?, *Electronics*, 65, p. 11, 15 June 1992.

29. Riley, W.D., Daellenbach, C.B., and Gabler, R.C., Jr. Recycling of Electronic Scrap, *Metals Handbook*, 10th ed., Vol. 2, *Properties and Selection: Nonferrous Alloys and Special-Purpose Materials*, pp. 53-61, 1990.

30. Sum, E.Y.L. The Recovery of Metals from Electronic Scrap, *JOM*, 43(4), pp. 53–61, April 1991.

31. Hoffman, J.E. Recovering Precious Metals From Electronic Scrap, *JOM*, 44(7), pp. 43–48, July 1992.

32. Reinhardt, A., Perratore, E., Redfern, A., and Malloy, R. The Greening of Computers, *Byte*. pp. 147-158, 1994.

33. Dambrt, S.M. NEC Develops New Printed Circuit Board Recycling Process, *Electronics*, 65, pp. 5–14, December 1992.

34. Ashley, S. Designing for the Environment, *Mechanical Engineering*, pp. 52-55, March 1993.

35. Owen, J.V. Environmentally Conscious Manufacturing, *Manufacturing Engineering*, pp. 44–45, October 1993.

36. Boothroyd, G. and Alting, L. Design for Assembly and Disassembly, *Annals of the CIRP*, 41(2), pp. 625–636, 1992.

37. Dewhurst, P. Product Design for Manufacture: Design for Disassembly, *Industrial Engineering*, 25, pp. 26–28, September 1992.

38. Wilder, R.V. Designing for Disassembly; Durable Goods Makers in Recyclability, *Modern Plastics*, 67, pp. 16-17, November, 1990.

39. Hundal, M.S. DFE: Current Status and Challenges for the Future, *ASME National Design Engineering Conference*, Chicago, DE VOL-67, pp. 89–98, March 1994.

40. "Tog-L-Loc," BTM Corp., Marysville, MI, *DesignFax*, ad, p. 174, February 1994.

7.10   PROBLEMS / CASE STUDIES

7.1   Automotive Seat Cushions: Johnson Controls Automotive Systems Group is incorporating design for disassembly into their automotive seats and interior components. The case study in question is the design of seat cushions. Two materials under study are PUR (polyurethane) and PET (polyethylene terephthalate). PET is one of the highest volume plastics used in manufacturing. Both materials can be recycled, but the PET is sought after since it is used much more, although PET cushion manufacturing is still under development and capacity is not in place.

Another design under development is the reprocessing of interior components. If materials, like polyester, can be cleaned, melted down, spun into fiber, and molded back into useful form, very little value is lost from the original product within this process.

a.   What other materials on the market could be used to replace the highly desired PET? Some items to consider, but not limited to these, when naming materials are:

- Is there a recycling program in place that would increase the reusability of these materials?
- Cost of virgin as compared to recycled raw materials.
- Are the materials compatible to the environment of the vehicle and other materials within the vehicle?
- Can the materials be easily incorporated into the cushion manufacturing or do they take extensive research?

b.   Think of other reasons why the materials chosen would be comparable or better than PET for the seat cushion.

7.2   Automotive Design: A 14-week senior design project at the Art Center College of Design in Pasadena, California, sponsored by the Automotive Applications Committee of the American Iron and Steel Institute illustrated how lighter weight production vehicles can be designed and manufactured with less glass, plastic, and fewer parts, demonstrating design for disassembly. Among the designs there were several features of the design for disassembly strategy. One such design was the use of stainless steel exterior panels that have gradual curves to reduce tooling expenditures and recycling value. Other designs included the use of adhesive and Velcro-like bonding systems that would speed assembly and more importantly disassembly. An important feature that nearly all the designs possessed was the clever elimination of plastic facials and body side moldings. Most important of all, the design project demonstrated that style and quality do not have to be neglected for the design of disassembly.

a.   Is there a conflict between designing for assembly and designing for disassembly?

b.   Develop several of your own strategies that incorporate the concept of design for disassembly. You may examine the life cycle of any product and categorize your guidelines into usage, disposal or recycling, and distribution.

c.   Select at least three automakers (e.g., Ford, Daimler-Benz, and Volvo) and describe their DFD and recycling efforts and determine the components that are currently being disassembled and/or recycled.

7.3   German Green Dot Program: This government-enacted program was set up with the philosophy that industry should be faced with the problem of "taking back" its products after they reach the end of their useful life span. This green dot program went into effect January 1993, and unfortunately, almost no data have been collected on the resulting impact on packaging consumption. There has been a great impact on public awareness of the environmental impact of waste disposal.

One of the lessons that can be learned from this program is that much of its success will depend on the cost-effectiveness of the various recycling programs, for example, plastic recycling. Unfortunately, in today's market, the cost of recycled plastic resin exceeds the cost of all virgin resins with the exception of PET. With the green dot program in effect, Germany now has stockpiles of recyclable plastic. Also, consumers in Germany offered more material for recycling than could be collected.

Although this information sounds negative, this is only the recycling of plastics. The green dot program has set some optimistic goals on the percentage of recycling metals that should be achievable. Also, there is an infrastructure in place to deal with the recycling of products that many other countries have not incorporated yet.

a.   List several products that are on the market today where the manufacturer could collect the product at the end of the life cycle and can easily recycle it. Cite your reasoning or thought process for choosing the products that you chose to include and drawbacks that may be present in the current design of the products.

b.   Of the products that you chose containing drawbacks, what would you recommend be changed in the design of the product to enable it to be more easily recycled?

7.4   Milk Carton Dilemma: A certain milk company is interested in changing its glass container to a nonglass container. The company wants you to come up with an economical yet practical design. Use AD methods and list the functional requirements and constraints. Then brainstorm, come up with ideas that may satisfy these requirements. After choosing a design, state tests that can be done to confirm the satisfaction of the requirements and constraints. Then state what the next few steps of the design would be.

7.5   Electric Outlet Design: Most of electric outlets have design as shown in Figure P7.5. This design consists of the following components:

- Base rivet
- Plastic base
- Power screws
- Power conductor
- Ground screw
- Ground conductor
- Conductor
- Conductor rivet
- Wall screw
- Plastic cover
- Cover rivet

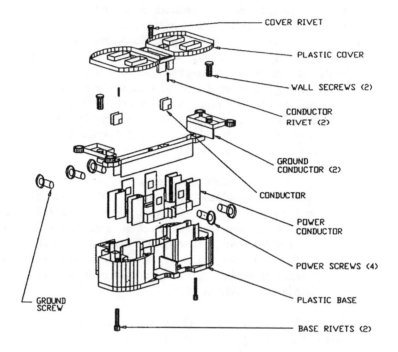

Figure P7.5   Original design of electric outlet.

  a. List components by name and identify the function of each.
  b. Perform analysis on all components and show how they work together.
  c. Conduct a DFA/DFD analysis on the original design and determine the design efficiency.
  d. Modify this design and determine design efficiency of the modified design.

7.6 Electric Light Switch Design: Consider the electric light switch design as shown in Figure P7.6. This design consists of the following components:

  • Plastic base
  • Spring
  • Rubber stop
  • Metal conductor
  • Power screw

- Plastic switch
- Plastic front plate
- Metal ground conductor
- Ground conductor rivet
- Ground screw
- Wall screw

a. List components by name and identify the function of each.
b. Perform analysis on all components and show how they work together.
c. Conduct a DFA/DFD analysis on the original design and determine the design efficiency.
d. Modify this design and determine the design efficiency of the modified design.

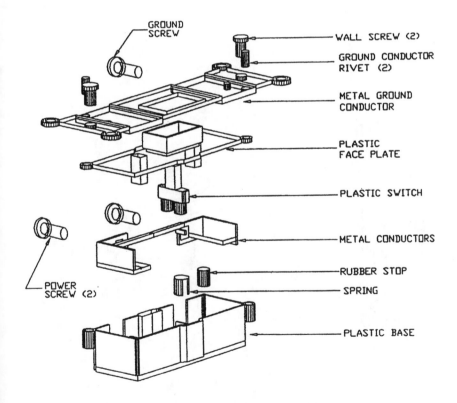

Figure P7.6 Original design of electric switch.

7.7   Water Filter Design: The original water filter design shown in Figure P7.7 consists of two subassemblies. Subassembly 1 consists of the following nine components:

- Valve assembly
- Top cover
- Screw ring
- Screws (3)
- Cartridge
- Cover

Subassembly 2 consists of the following four components:

- Body
- O-ring (two)
- Front O-ring
- Switch

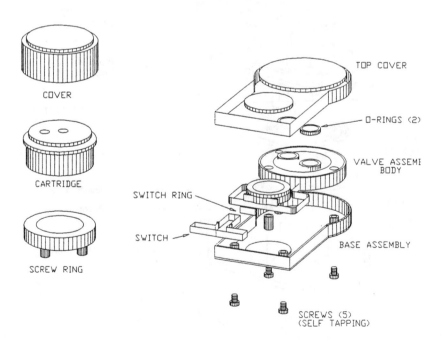

Figure P7.7   Original design of water filter.

a.   List components by name and identify the function of each.

b.   Perform analysis on all components and show how they work together.

c.   Conduct a DFA/DFD analysis on the original design and determine the design efficiency.

d.   Modify this design and determine the design efficiency of the modified design.

7.8   Consider a moderately complex product, such as a telephone. List all the materials required to make the product, and all the resources required to make the materials, including how much ore, oil, raw materials, water, energy, etc., is used and how much waste and pollution are generated. Report this breakdown with some descriptive graphics, such as flow diagrams, pie charts, etc. Where are the areas of greatest waste and what can be done to save resources?

7.9   Make a comprehensive comparison of many different fastening methods, such as threaded, adhesives, rivets, spot weld, snap fits, Tog-L-Loc, brazing, welding, etc. [39,40]. Compare the pros and cons for assembly time, cost, waste, pollution, reliability, and ease of disassembly for recycling. Summarize a set of rules that can be used for general design guidelines.

7.10   The Personal Computer: The personal computer has gone from an office curiosity to a business mainstay and common home appliance in a span of less than 15 years. The leaps in technological improvements have been even more rapid than their growth in numbers. An unfortunate consequence of the rapid advances in personal computer technology has been the realization that almost any new improvements are surpassed in 6 months. This in turn leads to equipment being rendered obsolete more rapidly than in any other industry, and hence, a very short life cycle. A recent study conducted at Carnegie-Mellon University estimates that if we continue to discard computers at the current rate of 10 million per year, there will be 150 million of them landfilled by the year 2005 [26].

a.   What is one to do with an obsolete or unrepairable PC? Donate it to a charitable organization? Heave it in the trash? Let it collect dust on the shelf? Or devise a way to reuse at least some of its components?

b.   Describe the different components, materials, and physical makeup of personal computers.

c.   How can one recover precious metals from the PC (e.g., silver and gold, with lesser amounts of platinum-group metals present [29]). Consider printed circuit board processing techniques to remove metals, such as hydrometallurgical or pyrometallurgical processing.

d.   Personal computers contain printed circuit boards that consist of glass fiber, layered epoxy resins, and copper traces. Are there ways to reclaim these components?

e. The monitor for a PC consists of two major parts: the plastic case and the CRT. As discussed earlier, the plastics can be easily recycled, but the CRT is a bit more troublesome. The problem is that CRTs contain lead to shield the viewers from the harmful X-rays produced by the electron guns. How can recyclers deal with this problem?

f. One important alternative to disassembling a PC into spare parts and raw materials is refurbishment and reuse. Is it possible? How?

g. Finally, to help PC manufacturers achieve the necessary goals of easy disassembly and recyclability for PCS, some DFA/ DFD guidelines should be implemented. Provide a practical list of these guidelines.

EIGHT

DESIGN FOR ENVIRONMENTALLY
FRIENDLY VEHICLES

This chapter provides a discussion of the current technologies used to minimize the detrimental impact that the automotive manufacturing industry has on the environment. It focuses on the change from the highly polluting internal combustion engines to the use of a cleaner, more efficient means of transportation, the electric vehicle. It deals with the major obstacles facing electric vehicles today. The chapter also examines the existing methods of improving the environmental aspects of the tires used by the automotive manufacturing industry. To achieve this goal, tires can be produced from recycled materials or recyclable materials.

8.1   INTRODUCTION

Warren Leon, Deputy Director for Programs at the Union of Concerned Scientists, once said that "We should use the market to encourage environmentally sound decisions.... We need to help the public and policymakers focus on those areas that would have the greatest impact." This environmentally conscious concern and many others have recently caused legislation to be passed that encourages the development of zero emission vehicles. The desire to improve air quality and minimize pollution from internal combustion engines is obviously the major push behind the development of the zero emission vehicle. Designers, however, not only need to eliminate the emissions from these vehicles, but they also have to make them viable alternatives to the cars that consumers already take for granted.

As engineers begin to tackle the problem of designing zero-emission cars, it is important that they take into account the overall environmental impact the vehicle will have. For example, an electric vehicle (EV) may be powered by batteries that need to be replaced every 2 years. Improper disposal of these batteries may lead to soil and groundwater contamination. In their attempt to design the environmentally friendly vehicles of tomorrow, engineers must be careful to look into all areas of the environment that might be impacted, even if it is through misuse of the vehicle.

For more than a decade now, people have been taking a serious look at the ways in which modern society affects the environment [1–8]. Governments around the world are responding to this heightened concern for the environment by passing legislation requiring both industry and the private individual to limit their impact on nature. One example of this is the push to develop alternative fuel sources for automobiles with the hopes of both eliminating the air pollution created by gasoline-powered vehicles and diminishing our dependency on fossil fuels.

The U.S. Environmental Protection Agency (EPA) estimates that more than 50% of all the air toxins in the United States are produced by motor vehicles [9]. The Alternative Motor Fuels Act of 1988 and the Clean Air Act Amendments of 1990 provided incentives for the U.S. automotive industries to build alternative fuel vehicles that would not contribute so much as to air pollution [8,10]. California has gone even a step further. The California Air Resource Board requires that, by 1998, 2% of all vehicles sold by a manufacturer must have zero emissions. Other states, including New York and Massachusetts, are following California's lead.

Detroit and environmental regulators have opposing viewpoints concerning electric vehicles [11–15]. Automakers fear that the technology necessary to produce cars that can compete with current internal combustion powered vehicles is still over a decade away. Executives representing Detroit's Big Three automakers are concerned that the technology is too primitive to produce electric vehicles that will be both radical and affordable for consumers. They maintain that car companies will suffer economically as a result. Major automakers, therefore, are trying to convince the EPA to ease zero-emissions regulations in an effort to mitigate the financial losses. Environmental regulators, on the other hand, argue that Detroit has a vested economic interest in preserving the market on conventional vehicles [16].

The automotive industry has attempted to eliminate or at least extend the deadlines imposed on it by the new legislation. Many companies expect to take huge losses when the first generation of environmentally friendly vehicles hits the market.

8.2   ALTERNATIVE FUELS

Potential sources of fuel for new and zero-emission vehicles can be broken down into two categories: renewable and nonrenewable. Nonrenewable sources, such as fossil fuels, are found in limited quantities in the earth [3]. Renewable sources can be continuously produced, like solar energy. Nonrenewable sources currently have an economic advantage over renewable sources, as evidenced in the following sections:

8.2.1   Renewable Sources of Power

Methanol. Methanol is produced from coal, natural gas, heavy oil, wood, and from methane produced from municipal waste. Some members of the U.S. automotive industry are backing clean-burning methanol as the most feasible alternative fuel source, primarily because of the large coal and natural gas reserves in the United States. Problems with methanol include its toxicity, driving range, corrosive characteristics, and ozone-creating formaldehyde emissions. Methanol is also 100% miscible with water, creating an environmental risk to water life in the case of a spill [17].

Liquefied Petroleum Gas (LPG). LPG is a mixture of gaseous hydrocarbons that are liquefied by pressure and/or reduced temperature. LPG is a high-octane fuel that reportedly produces lower levels of hydrocarbons than gasoline. Automobiles powered by fast-burning LPG would have less range than their gas-powered counterparts and would suffer from reduced cargo space due to the large canisters required to store the fuel.

Compressed Natural Gas (CNG). CNG is favored by the U.S. automobile industry and the U.S. petroleum industry because natural gas is abundant in the United States and is less costly to produce than gasoline. Vehicles powered by CNG produce nitrous oxide, a low atmosphere ozone-forming pollutant.

8.2.2   Nonrenewable Sources of Power

Ethanol. Ethanol is produced principally from corn and sugarcane, although it can be produced from nonrenewable sources like petroleum. Twenty-five percent of all cars manufactured in Brazil run on ethanol made from local sugarcane.

Hydrogen. Hydrogen-powered vehicles are virtually pollution free, emitting mostly water vapor along with small amounts of nitric oxides. If hydrogen is obtained from water through electrolysis, this fuel would be inexhaustible. Hydrogen fuel cells have been successfully used to power busses. However, the expense and danger associated with the distribution and storage of liquid hydrogen make this alternative fuel less viable commercially. Also, at present, hydrogen-powered vehicles have shorter range, less power, and 10–15 minute filling times as opposed to the conventional gasoline-powered vehicles [18].

8.2.3   Electric Power

Electric-powered vehicles can be considered either renewable or nonrenewable, depending on how the electricity powering the vehicle was produced. A car recharged with power from a hydroelectric plant would be making use of a renewable resource. A car powered with energy from a coal plant would be using a nonrenewable source of energy [19, 20].

Electric vehicles are the only automobiles that can currently meet California's zero-emission standards [21–24]. The hydrogen fuel cell could meet the criteria, but the difficulty involved in delivering the hydrogen makes it unrealistic to implement in the near future. All of the other forms of alternative fuels listed are considered to be economically unsound, too dangerous, or too unfriendly to the environment to pursue. For these reasons, the automotive industry has decided to promote the electric car as the environmentally friendly car of the future.

8.3   THE HISTORY OF ELECTRIC VEHICLES

With all the extravagant battery research and high-tech electronics being proposed for the electric vehicles of tomorrow, it is hard to believe that the first electric cars were driving at the turn of the century. Originally, there were three different technologies competing to power the horseless carriage: electricity, steam, and internal combustion.

Battery-powered cars, not surprisingly, took an early lead in the race to provide a reliable means of transportation. After all, cars were quiet, easy to control, and produced no odorous emissions. The science of storing electricity was also well understood at the time. The land speed record was actually held by an electric car in 1899 with an average speed of 40 miles per hour.

When Henry Ford perfected his gasoline-powered internal combustion engine, electric vehicles began to lose favor. In 1912, there was a major push to standardize electric vehicles in order to implement charging stations. However, the short range of the electric-powered cars and the lack of charging stations eventually led to the electric car's demise and to the rise of the gasoline-powered car.

8.4   ELECTRIC VEHICLES TODAY

Electric vehicles must overcome several major obstacles before they can compete with today's gasoline-powered vehicles. Range is one major problem. It will be difficult to convince consumers who currently travel up to 400 miles on a tank of gas that a car with a range of only 70 miles will serve their needs. Seven-hour recharge times are also a major problem. The new technology of pulse charging can charge the batteries in only 15 minutes, and plans to equip parking lots with special spots for electric vehicles can help increase the electric market. Electric vehicle manufacturers also plan to advertise electric vehicles as an ideal second car for short trips around town.

Another problem also related to the batteries is their limited life. Charging and discharging the battery packs can burn them out in as little as 2 years. Pulse charging has been found to extend their life, but the problem is still not eliminated. The economic impact on the owner and the potential environmental impact resulting from disposal problems make this a problem that should not be overlooked.

Initially, the cost of electric vehicles will be quite a bit higher than conventional vehicles because of the low-volume production. Tax breaks and selling to fleet buyers, like utilities, will increase the initial sales and help bring down costs.

8.5   ELECTRIC VEHICLE TECHNOLOGY

Car manufacturers hope to draw on the latest technological advances in such areas as energy storage systems, composite materials, and electric motors in order to produce their electric vehicles. Perhaps the greatest challenge is presented by the problem of energy storage. When one considers the fact that two conventional lead acid batteries contain only as much energy as a shot glass of gasoline, it is not hard to understand automakers' concerns over consumer reaction to the introduction of electric vehicles.

Many of the problems facing electric vehicles stem more from public misconception than from a lack of technology. The technology does exist today to manufacture excellent electric vehicles; the problem is that the technology necessary for electric vehicles to compete against gasoline-powered vehicles does not exist or may never exist.

Based on the Ford Ecostar shown in Figure 8.1, an example of a typical electric minivan contains:

- Tires: inflated to 65 psi to reduce rolling friction.
- Battery pack: Sodium sulfur batteries weigh 780 lbs.
- Electronics: Converts DC current to three-phase AC current.
- Motor: 75-horsepower, AC induction motor.

Characteristics.
- Maximum speed is 70 mph.
- Range is 80–100 miles before recharge.
- Recharge time: Between 6 and 7 hr.

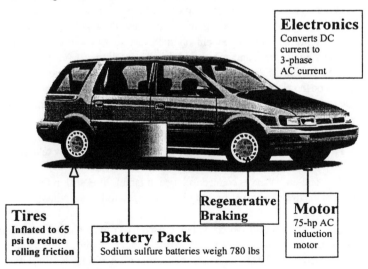

**Range: 80-100 miles Max. speed: 70 mph 0-60: 16.5 seconds
Recharge time: 6-7 hours**

Figure 8.1 Typical electric minivan.

Motors. Originally, electric vehicles were driven by DC motors. New developments in microelectronics now make it possible to convert the DC current stored within the batteries to three-phase AC electric current. AC motors are less expensive, smaller, more efficient, and more reliable than DC motors. AC motors are also better in taking advantage of regenerative braking, the process of capturing the energy lost while braking and converting it back into stored energy in the battery.

Bodies and Frames. Most of the first-generation electric vehicles will be little more than a gasoline-powered vehicle's body with batteries strapped to it. Engineers look forward to one day designing car bodies specifically for electric power. These bodies will be manufactured out of the latest composite materials to make them lighter to extend battery life. Aluminum frames might also be employed to save weight.

Power Supply. Perhaps the biggest obstacle facing the development of the electric car is the method to store the energy used by the motor. The problem is considered to be so severe that the Big Three automakers have teamed up with the U.S. government to create a U.S. Advanced Battery Consortium (USABC). The USABC hopes to develop a battery that can be mass produced and will allow an electric vehicle to perform as well as a gasoline-powered vehicle. Many small companies and university researchers have also attempted to solve the problem of storing power for environmentally friendly vehicles.

Chemical batteries release power through chemical reactions that take place with the battery. The chemicals within the battery deteriorate with each charge and discharge cycle, limiting their useful life. Many different types of batteries making use of various chemical combinations are being investigated as possible candidates. Figure 8.2 summarizes the performance capabilities of some of the batteries being considered. The data included in the figure are only a best guess of actual performance. The data are dependent on such factors as the number of batteries, weight of the vehicles, average speed, and driving conditions.

The lithium-polymer battery is being developed by the USABC. It is considered to be the electric vehicle's best bet to overtake gasoline-powered vehicles. The problem is that the ability to mass produce the batteries is still a major obstacle. Each type of vehicle, truck, minivan, sports car, sedan, etc., has a different structural frame. This has created difficulty in using one (or a few) battery design(s) in every vehicle. Therefore, mass producing batteries is a problem that may take many years to solve [13].

The zinc-air battery is being developed in Israel. It is still in the early testing stages and has the drawback that part of the battery has to be periodically replaced to refuel it.

The sodium-sulfur battery is being used by Ford in its Ecostar Minivan in order to extend the range. It is a very costly battery, contains materials that are potentially dangerous, and has a high running temperature of 650°F. Ford was forced to temporarily park all of its Ecostar prototypes when two caught on fire during recharging.

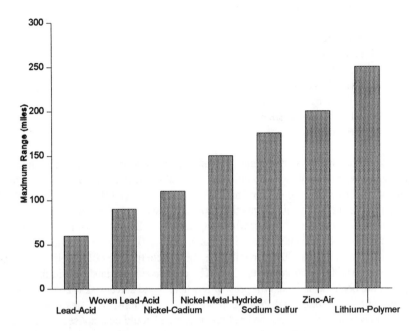

Figure 8.2. Maximum travel distance on one charge (based on an 18,000-lb battery pack; note that values are approximate).

Nickel-metal hydride batteries are being developed by a small company in Michigan with aid from the USABC. These batteries are currently being used in laptop computers, and the challenge is to make scaled up versions that can be mass produced economically. The batteries boast more environmentally friendly chemical compositions than the other alternatives and a fairly long life. The automotive industry is not overly optimistic that these batteries will be ready in time to power their electric vehicles at an affordable cost, but they still do not want to give up hope entirely on this promising battery.

Japanese automakers are looking at reliable nickel-cadmium batteries to power their cars. These rechargeable batteries have been used in toys and electronics and have quick recharge cycles, long lives, and a large capacity. The use of costly and environmentally dangerous cadmium makes these batteries less attractive to American companies.

Ovanic Battery Company, a subsidiary of Energy Conversion Devices, is one of the companies receiving research funding from the USABC. It has developed a nickel-metal hydride (Ni-MH) battery for use in any electric vehicle. During March 1994, the Ovanic Battery Company entered into agreement with General Motors to continue the development of this battery. Under this agreement, GM will build a production facility to mass produce the Ni-MH battery beginning in 1996. In a 1994 staff report, the USABC cited that the Ovanic Battery technology is projected to meet or exceed all of its midterm performance and cost goals by 1998 [25]. The primary battery performance criteria include:

- Specific energy, Wh/kg
- Energy density, Wh/liter
- Specific power, W/kg
- Cycle life (cycles)
- Life (years)
- Ultimate price, $/kWh
- Recharge time

The overall performance characteristics address many of the limitations associated with EV batteries. Specifically, Ovanic has demonstrated an on road range capability of over 150 miles using an 18-kWh battery and a 60% recharge capability in 15 minutes [25]. Ovanic claims that replacing the current lead acid battery, which powers GM's two-seat Impact, with an Ovanic Ni-MH battery of the same weight would increase the vehicle's range to approximately 250 miles. Assuming the battery is recharged to around 80% capacity, the 1000 cycle lifetimes of the Ni-MH battery equate to a 200,000-mile battery life. The electrode materials used in the Ni-MH battery are nontoxic and thus will present few health and safety risks [26]. According to the U.S. Environmental Protection Agency regulations, batteries using Ni-MH technology would not be classified as toxic waste and could be disposed of in landfills [25]. Ovanic's Ni-MH battery technology is also finding applications in other commercial devices such as portable computers and cellular phones.

The Horizon battery, produced by Electrosource, is an advanced lead-acid battery and is likely to be found in some of the electric vehicles sold in 1998 and beyond. Electrosource claims to be able to achieve ranges of up to 140 miles between charges and a 50% recharge in less than 8 minutes [27]. Lead-acid technology was used to power the first EV at the turn of the century, as is used today in automotive batteries. The fact that this technology is so well established provides a head start on newer battery technologies, which are just past their infant stages, and thus more expensive. Unfortunately, very few advances have occurred in this area until the impetus provided by the zero-emission vehicle mandates. Lead-acid technology simply did not perform well enough to be used in electric vehicles until Electrosource developed its horizon battery.

Although these batteries go a long way in solving some of the performance shortfalls suffered by EVs, the prohibitive cost still needs to be addressed. The $87,000 premium on the Chrysler EV and the $46,000 price tag on the Ford battery occur in part because of the extremely small production volumes. High conversion costs including the costs of internal combustion engine parts that are not needed is another common cost multiplier. When EV production volume is increased to at least 20,000 units annually, the body, chassis, and drive train will cost no more than their gasoline counterparts [28]. This leaves the cost of the batteries to determine the difference in price between electric vehicles and gasoline vehicles. One study indicates that by 1998, a lead-acid battery with a specific energy of 40–50 Wh/kg will cost $100–150/kWh when produced in large volumes. This equates to a battery price of around $1500 for a car with a 14–15 kWh battery pack such as the Geo Prizmbased sedan sold by U.S. Electricar [28].

Lifetime cost comparison is a better indicator of ultimate cost than the initial purchase price of the car. Table 8.1 illustrates the operating costs of EVs versus gasoline vehicles over a 10-year life [28]. The overall operating cost for an EV could even be lower when the reduced maintenance costs of the simple drive trains and fewer moving parts offered by EVs are considered.

It is still not surprising that the Big Three auto manufacturers balk at the zero-emission vehicle mandates. They have been traditionally resistant to change and are the world's largest producers of internal combustion engines. They are also being forced to sell an expensive product to consumers who may not want to purchase it. However, the number of impending mandates could approach 15 or more by the time California's goes into effect. The worldwide demand for EVs is also growing at least as fast as that in the United States, and the Big Three should look at this opportunity to increase their export market share. The opposite approach will lead to market share erosion as EVs produced in Japan, Germany, or France meet the world demand.

Many of the first electric vehicles to roll off the assembly line will make use of common lead-acid batteries found under car hoods today. These batteries are cheap and have a long proven track record. Improvements to these batteries, such as using a woven screen of lead, can provide twice the capacity, twice the life, and a quarter of the recharge time. Lead-acid batteries are very heavy and also contain toxic lead.

Some of the disadvantages of batteries may be offset through the use of supercapacitors that could provide a boost of power to the motor when accelerating or climbing a hill. Regenerative braking is also being used to help conserve power within the automobile. Regenerative braking takes the energy that would normally be wasted when slowing down the car and stores it in the battery for later use.

Some researchers are very excited by the possibilities of using a mechanical means of storing electricity instead of chemical means. The use of

Table 8.1. Operating Costs of Electric vs. Gasoline Vehicles Over a 10-Year Life

	Electric	Gasoline
Fuel cost	$0.05/kWh[*]	$1.25/gal.
Vehicle efficiency (4-door sedan)[**]	3 miles/kWh (equiv. to 100 mpg.)	25 mpg.
Fuel cost per mile	$0.017/mile	$0.05/mile
Battery cost per mile	$0.0312/mile[***]	N/A
Total operating cost per mile	*$0.048/mile*	*$0.05/mile*

[*] National average off-peak rate. In addition to price, energy security is a benefit of electricity: Nationwide, only 4 % of electricity is generated from petroleum.

[**] Measured from the wall plug and the gas pump.

[***] Assumes a $1,500 battery with a life of 600 cycles; 80 miles/cycle.

composite materials makes it possible to store energy in a high-speed flywheel. These flywheels may spin at over 200,000 revolutions per minute and provide plenty of power to the car. Ranges as far as 280 miles are projected for this mechanical means of storage. The technology still needs to be developed further before it can be used as a car's sole source of power.

8.6 ENVIRONMENTAL PROBLEMS FACING ELECTRIC VEHICLES

It is all too easy to fall into the myth that electric vehicles will solve all of the pollution problems currently facing our environment. This, of course, is not the case. Electric vehicles will impact the environment differently than gasoline-powered vehicles. Engineers need to consider this fact when designing vehicles that they hope will clean up the environment.

Although electric vehicles do not emit any toxins into the atmosphere, many of the power stations that generate the electricity used to recharge the batteries do emit toxins. Figure 8.3 shows the distribution of power produced by the various types of power stations in the United States. Note that most of the power is generated by burning coal.

It is often argued by many electric car enthusiasts that it is easier to control emissions at stationary power plants than aboard moving vehicles. Figure 8.4 shows that transportation generates far more oxides of nitrogen than the electric utilities, and it would appear that eliminating or even cutting back on car emissions would greatly benefit the environment. Other air pollutants, such as sulfur dioxides and particulate matter, are generated mainly in stationary combustion processes like power plants.

It is also important to remember that as the number of electric vehicles in the United States increases, so will the demand for electricity. This may lead to even more

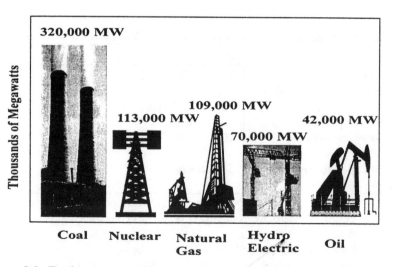

Figure 8.3 Total megawatts of power produced by power plants in the United States.

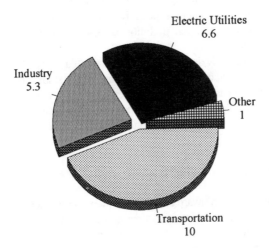

Figure 8.4. Estimated NO_x emissions in the United States (in 10^6 metric tons per year).

pollution unless cleaner burning fuels are used and the power plants use better filtration systems for the exhausted gases.

Coal has traditionally been the U.S. fuel of choice for producing electricity. The United States has huge coal reserves that can be taken advantage of to help diminish reliance on foreign countries for fuel. Coal, however, also produces more pollution than most other fuel sources.

Natural gas is a much cleaner burning fuel and is beginning to see more widespread use. Nuclear power plants release no pollutants into the atmosphere. Of course, the environmentally friendliness of nuclear power is the source of much debate.

The most productive renewable source of energy for generating electricity is water. Hydroelectric power is not feasible in all areas, and the environmental impact on wildlife in areas where water is being diverted to produce electricity can be extremely detrimental.

Other renewable sources of power generation are still many years away from being able to generate the power necessary to meet the demand. Geothermal power only has limited usefulness. Solar and wind power are still far from practical. One can also question the environmental soundness of covering acres of land with solar cells.

Power generation is not the only environmental issue facing the electric automobile that requires investigation. Every battery system being proposed for use in electric cars suffers from a finite life. The battery pack must be replaced once its useful life is over. The battery packs within the cars weigh anywhere from half a ton to a ton. Recycling of the chemicals used in the batteries is not always possible. Disposal can be a major problem, especially with the toxic materials found in some batteries. Groundwater and soil pollution is possible if care is not taken when disposing of the batteries.

Another potentially dangerous problem is related to the electromagnetic field generated by the electrical components in the car. Many studies have been launched into possible health risks associated with long-term exposure to electromagnetic fields. These fields are not only associated with the cars; if parking lots are equipped with spots for recharging vehicles, electromagnetic fields will be present.

8.7   POSSIBLE SOLUTIONS

Engineers must be extremely aware of the consequences that their design choices will have during this potentially revolutionary time in the automobile industry. The switch from gas to electric power would represent the most sweeping change in the automotive industry in its entire history. It would be extremely counterproductive if in another hundred years engineers were faced with the terrible dilemma of creating new cars that do not damage the environment like electric cars.

Disposal of the electric car bodies and frames can also be a problem. Composite materials used in the car may be very difficult to recycle. Efforts should be made to recycle plastic and metal parts.

The problem of providing electric power for a growing fleet of electric vehicles without polluting the environment further is a major problem that may never be solved with engineering ingenuity or future technological advances. The invention of advanced solar cells that will charge the batteries in a car in only 10 seconds on even the cloudiest of days would be an ideal solution, but unlikely. One solution is the recent development of photovoltaic (PV) cells, also called solar cells. PV cells are devices that convert sunlight energy directly into electricity, instead of using a thermoelectric conversion process. These cells could be placed on top of an EV to provide a continuous "trickle" charge while in operation and assist in the recharging process when parked at a recharge station. The recharge station could also use these cells to provide the source of energy [29]. A change in attitude by the average car driver will go a long way in furthering the crusade of electric vehicles and limiting human impact on the environment. Marketing is the key to successfully changing the opinion of the general public.

Perhaps the answer is not in creating family cars that will travel hundreds of miles on one refill, but in creating reliable vehicles that will allow one to travel short distances around town. The EV would be ideal for business use that requires travel within the city limits. The distances are short and the use of EVs within the city would help to reduce the smog level. Long-distance travel could be reserved for an improved and environmentally friendly mass transit system. The solution to this problem may not lie in the hands of the engineers alone, but all of society might have to decide on the future of transportation.

Some of the more immediate problems, such as the disposal of battery packs, do require immediate attention by engineers if future difficulties are to be avoided. The fact that used batteries pose an environmental threat is not disputed, but until other technologies such as flywheels or hydrogen fuel cells are perfected, batteries are a necessary evil.

Perhaps the most direct threat to the environment from disposal of depleted battery packs would be from illegal dumping of the batteries. This problem could be a major concern if electric car owners are held responsible for the disposal of their own batteries. The auto industry might diminish the threat of illegal dumping if the batteries were leased to car owners instead of being sold with the car. The responsibility and cost of replacing the battery would be in the hands of the auto dealer and not the owner. The system could even be backed up by placing specially designed locks on the batteries that could only be removed by qualified battery replacement specialists who would see to the battery's proper disposal.

Simply making sure that the batteries are not illegally disposed of still does not eliminate the problem of battery disposal. As electric cars become more common, the volume of batteries that need to be disposed of grows. The ease and extent to which a battery can be recycled need to be considered when choosing a power system for a car. This criteria should be even more important than the battery's performance characteristics. Since the main reason for undertaking the task of developing electric vehicles is to improve the environment, it makes little sense to create a major disposal problem just to extend the range of electric vehicles. A solution to limiting the accumulation of depleted batteries has already been instituted by a small company in Michigan. The company recycles the batteries' polypropylene casings and produces splash shields used on vehicles to protect the engine and other components from mud, rocks, and rust. Ford Motor Company presently uses these splash shields on four of their popular model vehicles totaling approximately 325,000 vehicles per year ,which equates to about 2.3 million pounds of the recycled polypropylene.

Recycling of unwanted electric vehicle parts does not stop with batteries. It should also be a major concern of design engineers to design car frames and bodies that make use of recycled materials. Plastic body parts and aluminum car frames are two prime examples. The engineers at Ford Motor Company have already instituted programs that use fewer types of materials and more that are recyclable. They are also using materials that have been recycled to limit the amount of newly produced materials. An example is the use of recycled polyethylene terephthalate (PET) soft drink bottles for parts such as grille reinforcement panels, headliners, and luggage rack side rails. Owners might be encouraged to return their old vehicles to recycling centers by offering to pay them back for the value of the materials in the car. The expense of recycling the materials would be passed on to new car buyers.

The potential health threats caused by the electromagnetic fields generated by the car's electronics should not be overlooked. More research is definitely called for in this area before thousands of people spend their commuting hours surrounded by harmful magnetic fields. The solution might involve "insulating" the driver and the passenger from these fields. Another solution might involve the use of a degaussing station that cancels the existing electromagnetic field with another electromagnetic field. This process could be performed after a recharging evolution that causes a buildup of an electromagnetic field. This process is used today on the ships of the U.S. Navy due to the buildup of an electromagnetic field from the electricity that is produced on board when the ship is underway. The problem might be considered severe, but not to the extreme that the search for the ideal electric car should be canceled.

The future of electric transportation is drawing nearer and it is full of many promises. Cleaner air, less noise, and a healthier environment are the treasures sought after by the electric car enthusiasts. The idea is simple enough: Eliminate one of the major sources of pollution, the internal combustion engine, and all of the problems it causes will go away. However, the perfection of the electric car will not lead to the complete replacement of gasoline powered vehicles, nor will it eliminate all adverse effects transportation has on the environment.

The fact that electric vehicles are not the perfect vehicle does not mean that research should be halted. Electric vehicles very well might make an extremely positive impact on the environment. As the automobile industry begins to take steps in the direction of designing the vehicle of the future, it is important that we make certain that it is headed in the right direction.

8.8   THE FUTURE OF ELECTRIC VEHICLES

Electric-powered vehicles are currently considered to be the only feasible approach to meeting California's zero-emission laws. Many people look forward to the day when electric vehicles will own the road. These people envision a world free of smog and full of silent, energy-efficient vehicles.

Futurists imagine a world equipped with special parking spots that would allow people to recharge their car batteries while shopping. The convenience would be extended to people at work. There might even be emergency trucks equipped to quickly recharge stranded vehicles. Electric-powered mass transit might even become a reality, cutting down not only on road travel but also on air travel.   Major benefits of electric vehicles include:

- Dramatic improvement in air quality
- The halt of adverse climate change
- Reduced trade deficit
- Enhanced national security
- Reduced water-supply contaminants
- Reduced landfill inputs
- Economic growth in advanced technologies

These are by no means small claims, and much research has been done to determine the extent at which these benefits can be achieved.

8.9   IMPROVEMENT IN THE DESIGN AND MANUFACTURING OF TIRES

Present automobile tire construction involves the use of reinforced thermosetting elastomers. This construction technique has been necessary to meet numerous design requirements such as tensile strength, resilience, abrasion and tear resistance, gas impermeability, UV radiation resistance, and high traction [30]. This construction, however, represents almost a worst-case scenario for recyclability. Several different recycling materials are generally used, which must be separated from the base elastomer material. In addition, the base material consists of several different rubber compounds. Rubber has a high energy content, and is difficult to break down while efficiently recovering the energy. Burning is an inefficient method of energy recovery and produces large amounts of airborne pollutants.

This problem can be thought of as having two aspects. First, the existing stockpile of used tires in the United States is estimated at 3 billion [31]. Second, an additional 285 million scrap tires enter the waste stream annually [32]. Tires constitute about 1% of the total waste stream. Tires placed in landfills tend to rise to the surface over time, where they provide sufficient breeding grounds for mosquitoes and shelter for rodents. This has led to the creation of dedicated, above-ground tire dumps [31]. These huge collections of tires pose a significant health and safety risk if they catch fire. Also, tires degrade very slowly in these dumps, due in part to their intentional UV resistance. Clearly, stockpiling is not a viable solution to our tire problem.

8.9.1   Tire Design Background

The modern automobile tire design is complex. The present-day pneumatic tire is the result of nearly a century of evolution. Figure 8.5 is a simplified diagram of a typical passenger car radial tire construction.

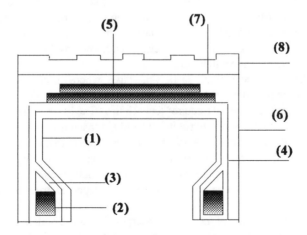

Figure 8.5. Typical tire construction.

Beginning from the inside, the synthetic rubber liner (1) provides the primary resistance to gas permeation. The bead's (2) wound steel hoops provide the strength and stiffness to keep the tire seated on the rim. The bead fillers (3) are wedge-shaped components above the bead to enhance the lateral stiffness of the tire and improve the load transfer between the bead and the sidewall areas. The plies (4) are fiber-reinforced layers that form the basic structure of the tire. The plies carry the stresses arising from internal pressurization, load, inertia, and tractive forces. The reinforcing materials used in the plies are usually polyester for auto tires and steel for truck/heavy duty tires. The belts (5) stiffen the tread area of the tire. The resulting reduction in deformation decreases rolling resistance and lowers tire operating temperature. Tire life is improved by reducing the amplitude of the cyclic strains in the plies, and by stabilizing the tread as it contacts the road surface. A material with a high elastic modulus is required for the belts. Steel is the most common belt material, though other materials, such as fiberglass, are used in less expensive tires. The sidewall (6) protects the plies from physical damage such as abrasion, cuts, and is compounded to resist UV radiation. The undertread (7) is compounded for flexibility and forms the base of the tread. The tread (8) is compounded for abrasion resistance, high traction, low noise, and UV resistance [25].

The percent by weight of each material in a typical automobile tire is as follows:

- Rubber—47.6%
- Carbon black—23.8%
- Chemical additives—14.3%
- Steel—9.5%
- Ply fiber—4.7%

8.9.2   Tire Manufacturing

Tire manufacturing has evolved into a relatively effective process, but is still quite labor-intensive. Many of the modern production engineering methods, such as in-process inspections and cellular equipment layouts, have been adopted. This has been necessary for U.S. firms to remain competitive with low-cost foreign manufacturers. The basic process is shown in Figure 8.6.

The process begins with Banbury machines, where the chemical constituents of the rubber are mixed under high heat and pressure. Several machines are used to produce the various compounds required for each tire. The batch is cooled into a slab, which passes through a series of rolling and cutting machines. The end products of these machines are the raw stock for the sidewalls, tread, and bead fillers. Simultaneously, the fabric for the plies is cut to size and coated with liquid rubber to aid adhesion to adjacent components. The beads and belts are similarly wound from steel wire and coated. The component materials are inspected prior to transferring to the next station.

The components of the tire are assembled on a tire building machine. The machine forms the components to their approximate shape and presses them firmly together. The end product of this step is called a "green tire." The green tire is

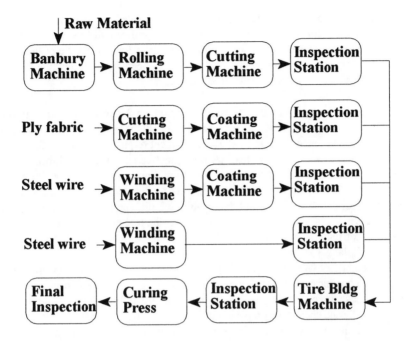

Figure 8.6. Present-day tire manufacturing process.

inspected prior to entering the next station—the curing press. The curing press presses the tire into its final form and vulcanizes the rubber. Vulcanization is the process of cross-linking the rubber molecules using sulfur and the application of heat. It is this process that gives the rubber its desirable properties, but also makes it difficult to recycle. The cycle time of this station is up to 25 minutes, depending on the size of the tire. A final inspection is done after curing [25].

8.9.3   Tire Recycling

Many methods are currently in use to dispose of used tires. These may include:

1. Burning for electric power generation.
2. Retreading.
3. Pyrolysis into basic constituents.
4. Use of ground rubber tire as an asphalt additive.
5. Combination of ground tires with thermoplastics.
6. Landfilling.
7. Stockpiling in tire dumps.
8. Undersea dumping to form artificial reefs.
9. Burning as fuel for cement kilns.

Of these, only items 1 though 5 fit the definition for recycling. These are each briefly discussed. Items 6 and 7 have already been ruled out as possible solutions. Item 8 does not consume enough tires to make a significant difference, and its long-term wisdom is questionable. Item 9 has been done for at least the past 20 years, though adequate air pollution controls have only been recently instituted, and are making this option less economically attractive.

1.   In the process of electric power generation, tires are burned to boil water, producing steam to operate a turbine generator. This is much like any other power plant. The difference is the specialized equipment required to feed the fuel into the burners, and to scrub the exhaust. A pilot plant constructed in California has the capacity to burn 5 million tires annually, providing power for 15,000 residential customers. This is a rate of about one tire/house/day. Problems have been encountered with the fuel delivery system and the scrubbers. Buildup of combustion products on the boilers is also a significant problem. However, the real problem is economics. The cost of electricity from the plant is $.085/kWh, five times the cost of electricity from fossil fuel plants [31].

2.   Retreading of worn tires is perhaps one of the oldest recycling concepts. It was done for economic reasons long before environmental consciousness became popular. The process involves stripping the remaining tread layer off an old tire, and mechanically and chemically preparing the surface. A new tread layer is then presses onto the surface and vulcanized. The process has limitations. First, the basic structure of the tire has a finite life expectancy due to fatigue and UV degradation. This generally limits the number of retreading cycles per tire to one. Second, not all discarded tires are suitable for retreading. Those discarded due to punctures, sidewall damage, or structural failure such as belt slippage are ineligible. Third, the modern complex radial tire is more difficult and expensive to retread than older designs. This, along with recent tire price depression, has removed much of the economic incentive for retreading of car tires [31]. The more durable and expensive steel reinforced structures of heavy truck tires make these economically viable retread candidates. Additionally, there is little market resistance to purchasing of retread truck tires, as there is for auto tires. For these reasons, retreading of truck tires is common.

3.   Pyrolysis is the process of breaking down the rubber components of tires into useful by-products. The process first requires the mechanical separation of the rubber from the fiber and steel reinforcing materials. The rubber is then heated in the absence of oxygen to prevent combustion. The heat breaks the carbon-carbon bonds in the molecules, releasing combustible gas, carbon, and oil similar in composition to crude oil [33]. The gas is used to fuel the burner, such that the process requires no external fuel source. The only pollution resulting is the combustion of the gas. Overall, the energy value of recovered materials is about four times that required to operate the reactor [33]. The carbon produces carbon black, suitable for reuse in the rubber manufacturing process. The oil can be used in rubber manufacturing or as fuel oil. Alternatively, the products may be used in the coal liquification process, where the carbon black is needed as a catalyst and the oil is used as a solvent [34].

4.   The use of scrap tire rubber that has received the most public attention is the addition of ground rubber tire (GRT) to bituminous concrete for road pavement. Bituminous concrete is a mixture of crushed stone aggregate and a binder called

asphalt. One reason rubberized asphalt is popular is the potential volume involved. It is estimated that if 25% of the bituminous concrete produced annually were rubberized, 200 million tires would be consumed in the process [31]. Rubberized bituminous concrete is quieter than ordinary road surfaces and provides an all-around better road surface. This new paving material is disliked by contractors due to the fact it costs up to twice as much as ordinary asphalt. Also, this asphalt lasts up to three times longer, which means paving contractors would lose a significant amount of business.

5. One of the most economically viable uses for scrap tires appears to be the addition of GRT to thermoplastics. The GRT can be added by simple melt blending with the base material. GRT improves the impact toughness of many polymers, including polyethylene, polystyrene, and polyurethane. The cost of GRT varies from $0.35 to $1.00 per pound versus base material costs of $1.25 to $5.00 per pound. Part of this advantage is that GRT production requires about 9000 Btus of energy per pound, including collection, whereas virgin polyurethane requires about 90,000 Btus per pound [31]. There are many markets for the thermoplastics with the addition of GRT:

a.   Agriculture—wheels for lawn and garden equipment, irrigation hoses
b.   Automotive—floor mats, mud flaps, bumpers, truck liners
c.   Construction—roof shingles, outdoor surfaces for track and field
d.   Footwear—shoe soles, inner sole foam
e.   Industrial—pallet skis, industrial wheels and housings

8.9.4   Present Initiatives

There are two major initiatives underway in the tire industry with potential positive environmental impacts. These are:

1. Development of low wear rate tire designs. The typical life expectancy of a modern automobile tire is about 40,000 miles. Several manufacturers have developed and tested designs with life expectancies of 100,000 miles. The higher mileages have been achieved through a combination of improved tread compounds and structural modifications to reduce cyclic elastic deformation. The structural modifications include higher inflation pressures, stiffer belts, and smaller tread blocks to reduce individual tread block strain. The harder tread compounds have some adverse effect on dry traction [35]. This is partially compensated for by the stiffer contact patch [36]. The structural stiffening has the secondary benefit of reducing rolling resistance for improved fuel economy. The reduced wear rate should reduce airborne tire tread products by a factor of 0.4. Airborne tire tread wear products have been identified as a significant pollutant in Tokyo [9].

There are some marketing problems with the long-life tires. The manufacturers are somewhat reluctant to commit production resources to the tires because of the resulting reduction in tire demand if the tires become commonplace. This would require marketing the tires at a higher price to maintain profitability. Additionally, the real-world life expectancy of a tire is very much a function of consumer attentiveness and maintenance. Many tires are discarded prematurely due to improper alignments, over- or underinflation, and harsh driving habits. Correction of even one of the aforementioned causes could substantially decrease the number of tires being discarded earlier than necessary. Further, a consumer expecting to keep an individual automobile for no more than 50,000 miles is unlikely to pay extra for 100,000-mile tires.

2. A secondary industry initiative is the run-flat pneumatic tire. The primary industry motivation for this is the desire of the automakers to eliminate the valuable space and weight (about 55 lbs.) of the spare tire and the jack [25]. The compact spare tires common in new cars were introduced as partial fulfillment of this goal. Tire manufacturers have been working on the run-flat designs for at least a decade. The primary problem with conventional tires upon loss of air is that the sidewalls buckle, allowing the rim to contact the road surface and the bead to separate from the rim [37]. The latest run-flat designs have an internal element that prevents complete collapse of the sidewall and bead separation upon loss of air pressure. Currently, this technology is available only on high-performance tires having a low sidewall height. The physics of the problem becomes more difficult as sidewall height increases [31]. The tires are capable of sustained 55 mph speeds when flat, equivalent to the compact spares [38].

The next steps required in the implementation of this technology are extensions to higher sidewalls, and effort to remove customer insecurity about the absence of the spare and jack. The environmental benefit is clear: Elimination of the spare tire would reduce the amount of tire rubber entering the waste stream by about 15%. This figure is based on a conventional spare tire constituting one-fifth of the tire rubber per auto and the fact that compact spares contain slightly less rubber than conventional tires.

8.9.5 Proposed Initiatives

Injection Molding Manufacturing Processes. The production of tires by injection molding has the potential to reduce the unit cost of tire production. The technology has been used for the production of heavy equipment tires since the 1960s, and at that time was considered and rejected for automobile tire production. At that time, tire companies had just made a large capital investment in equipment to produce radial tires and were not interested in changing their production methods again. However, the process may make economic sense in today's price-sensitive market.

A basic diagram of the process is shown in Figure 8.7. As compared to the current process shown in Figure 8.6, the reduction in the number of stations is apparent. Not as apparent from the figures is a reduction in materials handling equipment required. Both of these advantages stem from the fact that there is no raw rubber stock to be cooled, formed, and handled as an intermediate process. A corresponding reduction in manual labor would be expected. The process cycle time would still be governed by the time required for vulcanization. The vulcanization process could be expected to be slightly faster, because the liquid rubber entering the mold is closer to vulcanization temperature than to the green tire in the current process.

While the injection molding process itself has no environmental impact, it

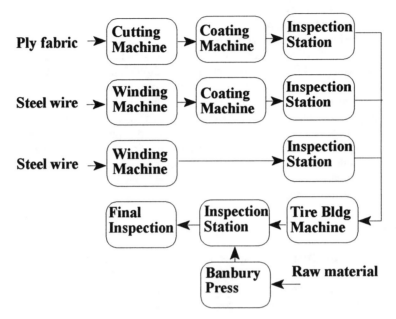

Figure 8.7. Injection-molding tire manufacturing process.

reduces production cost. This would allow one of two secondary initiatives: First, the process could be used to produce the long-wear tires at or near the cost of producing conventional tires by the current methods. Thus, the environmental benefits of the long-wear tires could be achieved without significant additional production costs. Similarly, the reduced production costs could be used by tire companies to incorporate tire reclamation processes, such as large-scale hydrolysis, into their production system.

Use of Thermoplastic Components. As thermoplastics continue to be developed, they are gradually replacing thermoset elastomers in many markets. To date, tire production is not one of those markets [39]. The raw material for thermoplastics is generally more costly than for elastomers. However, thermoplastics have two key advantages: decreased processing time and recyclability. Both of these advantages stem from the fact that thermoplastics do not require vulcanization. Thus, there is no curing time in the process between raw material and the finished product.

It is proposed that thermoplastics be introduced for specific components of conventional tires. There are some precedents for the use of nonelastomers in tires. Some compact spare tires are constructed of reinforced polyurethane [4]. The bead filler would be a logical first choice as a component for thermoplastic substitution. This component does not require abrasion resistance or traction. Thermoplastics are available that are suitably flexible and fatigue/temperature resistant. These materials can be surface modified to bond effectively to the base elastomer [40,41].

In the production system, the bead filler would likely be purchased as finished components from an outside supplier. One of the Banbury machines dedicated to producing bead filler compound could be eliminated or used to produce one of the other needed rubber compounds.

The bead filler is estimated to represent 5–10% of the rubber in a conventional tire, and therefore its replacement would cause a 5–10% source reduction. The thermoplastic bead filler should prove no more difficult to separate from the rubber and steel than the polyester ply reinforcement currently used.

This concept requires further evaluation from both technical and economical aspects prior to implementation, but would provide a wealth of experience among tire producers in dealing with thermoplastics. The desired outgrowth of this would be the gradual replacement of all rubber components in the tire, yielding a tire with only two components, steel and thermoplastic. These materials are readily separable and recyclable with current technology. Also, Ford and Michelin are currently teaming up to produce recycled tires made from scrap tires. These tires would cut about 30 million tires annually from entering landfills; this is only about 12% of the tires scrapped every year.

Nonpneumatic Designs. Nonpneumatic tire designs have the potential to produce two desirable effects: the elimination of spare tires, and increased use of easy to recycle thermoplastics. A nonpneumatic tire is currently under development in a project funded by the Swedish government. The tire consists of a steel hub, a reinforced thermoplastic wheel, and a rubber tread. The wheel is S-shaped in cross section and provides the necessary vertical flexibility, lateral stiffness, and deformation at the current patch. The design's drawbacks include greater noise transmission through the suspension, and a tendency to develop an out-of-round condition when parked for extended periods of time [36].

The potential environmental benefits of such a design should not be ignored. This design should be aggressively developed, to solve its current drawbacks, and to replace the rubber tread material with thermoplastic or other easily recyclable material. Even if implemented in its current form, it would result in an estimated 80% source reduction for tire rubber. Most of the remaining 20% would end up in the environment as treadwear products.

8.10 SUMMARY

Electric transportation is a new environmentally friendly venture that is full of many promises. Cleaner air, less noise, and a healthier environment are the treasures sought after by the electric vehicle enthusiasts. The concept is practical enough to implement that eliminating one of the major sources of pollution, the internal combustion engine, will eliminate or reduce other associated problems. To think, however, that the perfection of the electric vehicle will completely replace the gasoline-powered vehicle to save the environment is mistaken. The introduction of the electric vehicles also introduces the possibility of other environmental threats such as the battery disposal and the increased power plant output.

The fact that the electric vehicles are not the perfect vehicle does not mean research should be halted. Electric vehicles might very well affect the environment in the positive direction. As the automobiles industry takes the steps in the right direction of designing the vehicle of the future, it is important that society makes certain it is headed in the right direction. This direction must include all venues of design and manufacturing of the vehicle parts such as the body, the engine, the battery, the tires, etc.

As we enter the twenty-first century, it seems that the twentieth century has left us with much waste to be eliminated. Recycling and environmentally friendly design are becoming commonplace. With almost 300 million tires being scrapped annually in the United States, and 3–6 billion already stockpiled, it is imperative that recycling methods be explored and implemented. The supply certainly exists, but what is needed is the demand. Currently only 5% of all scrap tires are used in fabricated products.

Among the most popular methods of recycling tires are pyrolysis, burning for electric power generation, retreading, and the use of ground rubber as an additive to asphalt or thermoplastics. If left alone in stockpiles, tires are both a health hazard and an extreme fire hazard. Scrap tires are a serious problem, but if dealt with properly, they can be both very profitable and environmentally efficient.

8.11   BIBLIOGRAPHY

1.   Allen, M. Battery Chargers, *Popular Mechanics*, pp. 10–31, September 1991.
2.   Ashley, S. Aluminum Vehicle Breaks New Ground, *Mechanical Engineering*, pp. 50–51, February 1994.
3.   Ayres, R.U., and McKenna, R.P. *Alternatives to the Internal Combustion Engine*, Johns Hopkins University Press, Baltimore, 1972.
4.   Behne, T.A. LIM Plastic Tire Ready for Production, *Automotive Industries*, 164, pp. 50–51, March 1984.
5.   Brown, K. *UDI Who's Who at Electric Power Plants*, Utility Data Institute, 1992.
6.   Card, A.H., Jr. Technology Isn't Here Yet, *USA Today*, May 12, 1994.
7.   Collie, M.J. *Electric and Hybrid Vehicles*, Noyes Data Corporation, Park Ridge, NJ, 1979.
8.   Electric Hybrids Ready for LA, *Popular Science*, p. 32, December 199.
9.   Kim, M.G., Yagawa, K., Inoue, H., Lee, Y.K., and Shirai, T. Measurement of Tire Tread in Urban Air by Pyrolysis-Gas Chromatography with Flame Photometric Detection, *Atmospheric Environment, Part A, General Topics*, 24A (6), pp. 1417–1422, 1990.
10.   Flagan, R.C., and Seinfeld, J.H. *Fundamentals of Air Pollution Engineering*, Prentice Hall, Englewood Cliffs, NJ, 1988.
11.   Keebler, J. All Charged Up, *Automotive News*, p. 7, December 16, 1991.
12.   Kurylko, D. VW Quits Swatch Car Project, *Automotive News*, p. 7, January 25, 1993.
13.   Lave, L., Hendrickson, C., and McMichael, F. Environmental Implications of

Electric Cars, *Science*, pp. 993–995, May 19, 1995.

14. Perrin, N. In Defense of Electric Cars, *Audubon*, p. 18, September–October 1993.

15. Taylor, A. III. Why Electric Cars Make No Sense, *Fortune*, pp. 126-127, July 26, 1993.

16. Karten, L. EV Test Market Is Driven by Utilities, *Electronics World*, pp. 44–45, March 1995.

17. Knott, D. Alternative Motor Fuels: A Slow Start Toward Wider Use, *Oil & Gas Journal*, pp. 25–28, February 20, 1995.

18. Winter, D. Is Hydrogen the Ultimate Fuel?, *WARD'S Auto World*, pp. 50–51, April 1995.

19. New Battery Electric Cars, *USA Today News View*, p. 13, August 1993.

20. O'Connor, L. GM Backs Nickel-Metal-Hydride Battery, *Mechanical Engineering*, p. 24, May 1994.

21. Raynor Cannastra Associates, *The World of Electric and Solar Vehicles*, Edison Electric Institute, 1993.

22. Running Interference, *Popular Science* , p. 64, September 1993.

23. Shulidiner, H. New Age EV's, *Popular Mechanics*, pp. 27–29, 102, September 1991.

24. Sperling, D., Gearing Up for Electric Cars, *Issues in Science and Technology*, pp. 33–41, Winter 1994/95.

25. Ovshinsky, S.R. "Ovanic Ni-MH Batteries for Portable and EV Applications," March 1, 1994.

26. "NREL Examines Environmental, Health and Safety Issues Concerning Nickel Metal Hybride Batteries," National Renewable Energy Laboratory, July 1994.

27. *Electrosource News*, Electrosource, Inc., September 1994.

28. Electric Power Research Institute, Technical Brief, "Electric Vehicles Need Not Be Expensive," June 1994.

29. Teitelbaum, T., Energy Advocates on Sustainability, *Gaining Ground*, Global Action & Information Network, Vol. 3, No. 1, Winter 1995.

30. Jost, K. Tire Design and Development, *Automotive Engineering*, 100, pp. 21–25, June 199

31. Wallace, J. All Tired Out, *Across the Board*, 27, pp. 24–29, November 1990. 2.

32. Pennisi, E. Rubber and Road, *Science News*, 141, pp. 24-29, March 7, 1992.

33. Mustafa, N. *Plastics Waste Management* , Marcel Dekker, New York, 1993.

34. Farcasiu, M. Another Use for Old Tires, *ChemTech*, 23, pp. 22–24, January 1994.

35. Grosch, K. Some Factors Influencing the Traction of Radial Play Tires, *Rubber Chemistry and Technology*, 57, pp. 889–907, November/December 1984.

36. Scott, D. The Frisbee Tire, *Popular Science*, pp. 32–33, March 1992.

37. Williams, W. Technology Comes to Tire Design, *Popular Mechanics*, pp. 98–99, April 1985.

38. Schroeder, D. The Run-Flat Eagle Tire, *Car and Driver*, 39(8), pp. 95-96.

39. Gagne, J. Elastomers in the 90's—A Mixed Outlook, *Chemical Week*, pp. 30-34, May 8, 1991.

40. Rajalingham, P., Sharpe, J., and Baker, W.E. Ground Rubber Tire/Thermoplastic Composites: Effect of Different Ground Rubber Tires, *Rubber Chemistry and Technology*, 66, pp. 664–677, September/October 1993.

41. Rajalingham, P., and Baker, W.E. The Role of Functional Polymers in Ground Rubber Tire- Polyethylene Composite, *Rubber Chemistry and Technology*, 65, pp. 908–916, November/December 1992.

42. Healy, J.R. Fire Spark Safety Investigation of Ford's Electric Vans, *USA Today*, July 27, 1994.

43. Keep Up the Pressure to Develop Clean Electric Cars, *USA Today*, May 12, 1994.

44. Merchant, A., and Petrich, M. Pyrolysis of Scrap Tires and Conversion of Chars to Activated Carbon, *AIChE Journal*, 39, pp. 1370–1376, August 1993.

45. Washington, F., The War on Waste—Tire Recycling in the Automobile Industry, *Ward's Auto World*, 31(9), p. 61, September 1995.

46. Woodruff, D., Armstrong, L., and Carey, J. Electric Cars *Business Week*, pp. 104–114, May 30, 1994.

47. Woodruff, D., Templeman, J., and Sander, N. Assault on Batteries, *Business Week*, p. 39, November 29, 1993.

48. Jost, K. Tire Materials and Construction, *Automotive Engineering*, 100, pp. 23–28, 1992.

8.12 PROBLEMS / CASE STUDIES

8.1 James Inglewood is a design engineer at the Gathers Battery Corporation (GBC), a company that specializes in producing lead acid batteries. GBC has just landed an exclusive contract with the Ford Motor Company to mass produce lead acid batteries for the first wave of electric vehicles. Ford has specified in the contract that the same battery will be utilized in each of their different models. Ford is going to produce an electric vehicle version of the Arrester, Taurus, Contour, and Escort. In addition, Ford wants to maintain low lead emissions during use. They also wanted suggestions on how to dispose of the batteries with minimal effect on the environment.

James has just received the technical drawings of the structural frames for the aforementioned electric vehicle. He must now begin the task of designing a lead acid battery with these constraints. Can you help James to respond to the following dilemmas that he is facing:

a. Where on the frame should James decide to put the batteries, accounting for the problems of battery weight, size, and weight distribution of the car (remember that all the internal combustion parts will be absent and replaced with an AC induction motor)?

b. Discuss the effects that the location of the batteries will have on the overall safety and efficiency of the vehicle.

c.   Four different models will be produced with the same batteries. What problems will arise in the battery design as a result, and how could James solve these problems?

d.   What are some suggestions to dispose of the lead acid batteries in an environmentally safe manner?

8.2   Jeremy Reeve has just been hired to work on a project at the Global Battery Company in San Diego, California. Jeremy's supervisor has requested that he research the various batteries available today, with the task of determining which one would be the most beneficial to use in the future on electric vehicles (10–15 years). In order to determine the optimal choice, Jeremy must consider the following constraints:

• Greater than 150 mile maximum travel distance on one charge.
• Environmentally safe during use and disposal.
• Ability to be mass produced.

a.   Make a list of possible battery choices and outline the pros and cons of each.

b.   Determine which batteries meet all the constraints mentioned.

c.   Which batteries present the greatest threat to the environment and the safety of humans? Offer some ways to potentially minimize or even eradicate these adverse effects.

d.   Which battery would you choose for the future and why?

8.3   The state of New York produces about 15 million scrap tires annually. You are proposing to pass a law requiring that 5% of all asphalt that is to be used on state roads must consist of recycled tire crumb. Knowing that this is both expensive and frowned upon by the contractors, how would you "sell" your proposal to the state and to the disgruntled contractors?

8.4   Companies such as Vandeley Industries uses recycled tire chips mixed with plastics to manufacture many products such as floor mats, bumpers, and boat dock guards. List 10 new ideas (not mentioned in this chapter) for products that could be manufactured from this mixture and discuss why this is both profitable and environmentally friendly. (Keep in mind that the ideas you come up with cannot be dependent on color since the products will most likely have to be black due to the recycling process.)

8.5   Dr. Harry McDonahue is a senior project engineer at Future Automotive Technologies Corporation (FATC) in Atlanta, Georgia. His division is responsible for researching topics that pertain to production, development, and improvements of the electric vehicle. He is presently supervising a research team comprised of two other engineers, Ben Smith and Rebecca Harold. The objective of the project is to produce a detailed study that provides an objective comparison of the emissions produced by both internal combustion engines as well as the emissions produced by power plants that support EV use.

Ben discovered during his research that electric vehicles have a myriad of environmentally friendly attributes that would help to minimize air pollution. He found that electric vehicles are 90% efficient. Due to their regenerative braking capabilities, electric vehicles can recapture up to 50% of the energy lost during braking. Another 6% of energy use is reduced due to the fact that no transmission is needed. Furthermore, he found that another 10% of energy is conserved since these vehicles do not use energy during idling and coasting.

Power plants, on the other hand, are not as efficient as the EVs themselves. The efficiencies of the power plants of today are only about 33% for those using oil, natural gas, and coal. An increase in this efficiency to 50% is expected as various ways to improve these plants are developed and implemented.

Rebecca discovered that internal combustion engines are only about 10–25% efficient.

Dr. McDonahue has requested that Rebecca and Ben collaborate in their research efforts and form a comparison between the internal combustion engine emissions and emissions from power plants caused by electrical vehicle use.

a.   What important aspects of their findings should they include in their comparison?

b.   What are some of the concerns that might arise as electric vehicles and thus power plant emissions become more prevalent in the future? As an engineer representing FATC, how would you convince a disgruntled group of people at a town meeting that the power plant in their community is a favorable idea for the environment? Also, how could you convince people to embrace more nuclear power plants as a clean source of energy?

c.   Assume a power plant efficiency of 50% and an oil refinery efficiency of 90%. What is the difference in efficiencies between the internal combustion engines and electric vehicles?

8.6   One drawback to the present designs of electric vehicles is the production of electromagnetic fields during the charging process and also prolonged use of the electric vehicle. You have been hired as a design engineer to eliminate the problem of prolonged exposure of passengers to electromagnetic radiation.

a.   What changes, if any, could be made to the present design of the EV that would decrease or eliminate exposure to electromagnetic radiation? (Assume the current EV design is the present-day vehicle design with internal combustion components replaced with electrical components for the EV).

b.   The concept of a degaussing station has been mentioned briefly in the chapter. Generate a design for a degaussing station that would be aesthetically acceptable to the public and safe for use.

c.   Besides a degaussing station, what other methods could be instituted to eliminate the electromagnetic fields that are produced?

NINE

DESIGN FOR MANUFACTURING PROCESS IMPROVEMENT

The manufacturing process can be an efficient and environmentally safe process with a little foresight. During the design phase of the plant, problems can be solved more easily than if they are discovered after the plant construction is completed. This chapter presents a few of the many processes that can become environmentally safe procedures that would not affect the price of producing a product. From safer cleaning systems and wastewater treatments to improved painting techniques and disposal of aerosol cans, these processes can become friendly to the environment.

9.1   INTRODUCTION

Many manufacturing processes can be improved to become more environmentally friendly in a cost-effective manner. The use of recycling, regeneration, filtration, or simply the elimination of some steps may lead to a friendlier process. Some industrial efforts are presented in the following sections [1–4].

9.1.1   Solvents Industry

Industry is now moving toward the elimination of harmful solvents and replacing them with environmentally safe cleaners [1]. For example, at Acme Composites, great strides in the conversion from harmful solvents to water-based cleaners have occurred since April 1993. It took a major cost to make the company look at the problem in depth. It was losing 55 gallons of methanol a week to evaporation, 55 gallons of methylene chloride a month, and 3500 pounds of CFC-113 a year. The decision was made to eliminate these cleaners and move to water-based cleaners.

By April 1993, Acme had eliminated its uses of all CFCs and methylene chloride. Methanol had decreased 95% and methyl ethyl keytone (MEK) use was down a little over 50%. Both MEK and methanol were phased out by the end of 1993.

With water-based cleaners, Acme's environmental record improved dramatically, and product quality had improved nearly 15%. The switch also proved to be very cost-effective: Acme was saving more than $41,000 per year. It is true that there is no single, drop-in alternative, but with the combined efforts of Acme and one of their vendors, they were able to come up with a cost-effective, environmentally safe alternative while increasing quality.

> *Substituting water-based, or aqueous and semi-aqueous, cleaners for chloride solvents requires rethinking of the manufacturing process....You have to look at what you are doing. You have to modify practices, conduct tests, and measure results.*—Richard Lanier, Acme Supervisor of Manufacturing Engineering

With new products, the capitol costs associated with them often come under scrutiny, and are often deterrent to making the switch. In Acme's case, the new equipment was not overly sophisticated, nor pricy. Lanier installed the industrial equivalent of a home dishwasher to clean some parts.

9.1.2   Alternative Cleaning Costs

The cost of alternative cleaning solutions is deceiving when taken at face value. Several factors as shown in Figure 9.1 must be looked at to get the total picture [1–4].

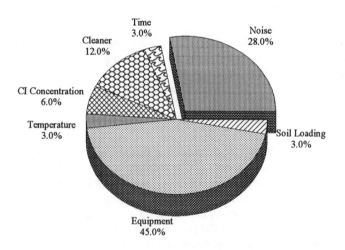

Figure 9.1.. Factors affecting cleaning.

An example of this is MEK costs about $3 a gallon, whereas the alternatives may cast $30 a gallon. After looking at a drum with a dilution of about 5%, the actual cost of the cleaner can cost less than half of the old solvent. Also with new cleaners, recycling can recover 80–90% of the product for reuse. This can be seen in the cost of removal. Referring back to the case of Acme, it is saving 80–90% on shipping and disposal costs due to the reduced volume of waste.

In other plating and coating applications, recycling and regeneration have been the key to reducing waste generation and the need for wastewater treatment. Closed-loop systems, deionization, distillation, and diffusion dialysis are all methods being employed to reduce the use of toxic chemicals. These methods also reduce the amount of chemicals required in the manufacturing process. For example, if a cleaning tank is filled and the contents are run through a distillation process and 50% of the original solution is salvaged, then only 50% more of the chemical needs to be added. This is a very simplified example, but these processes are very effective and are being looked at more and more.

9.2   ENVIRONMENTALLY SAFE PRECISION CLEANING SYSTEMS

The most common method of precision cleaning used in industry today is vapor degreasing. Vapor degreasers typically use chlorinated solvents to remove water-insoluble soils from the surface of a workpiece. Water-insoluble soils include grease, oils, waxes, tars, fluxes, and carbon deposits. These deposits must be removed from a workpiece prior to manufacturing processes such as machining, plating, painting, assembly, repair, inspection, or heat treatment. Further, vapor degreasing systems can reduce contamination in downstream processes such as electroplating.

Nonchlorinated solvents, such as mineral spirits and alcohols, can be used in solvent degreasing systems. However, chlorinated solvents have remained popular as degreasing solvents due to their solvency for organic materials, low latent heat of vaporization, nonflammability, noncorrosiveness, and relatively high stability. The most commonly used chlorinated solvents used in degreasing systems are [5]:

- Methylene chlorine (dichloromethane or DCM)
- Perchloroethylene (PCE or tetrachloroethylene)
- Trichloroethylene (TCE)
- 1,1,1-Trichloroethane (TCA, methyl chloroform, or MCF)
- Trichlorotrifluoroethane (chlorofluorocarbon 113, or CFC-113)

Unfortunately for industry, the production and use of the chlorinated solvents just listed will be restricted due to the fact that these chemicals contribute to the depletion of the earth's ozone layer [6]. Therefore, alternative methods of precision cleaning must be developed to satisfy the needs of industry worldwide. The alternative methods should replace the chlorinated solvents with solvents that do not have a detrimental effect on the environment.

9.2.1   Why Precision Cleaning?

This is especially true for technologically advanced components used in aerospace, defense, electronics, and medical industries. As the products being developed by these industries continue to evolve, the tolerances and surface finish requirements tend to become more stringent. Therefore, the requirement for a contaminant-free surface prior to operations such as finish machining, plating, and assembly cannot be overemphasized.

Machining lubricants and metallic residue are among the primary sources of contamination. Most of the conventional machining processes such as drilling, milling, and turning will generate a significant amount of heat during the fabrication of a part. The heat generated will tend to "bake" the cutting fluid on the surface of the workpiece. As a result, the workpiece becomes sticky and water-insoluble contaminant, such as metallic residue, will adhere to the part. The most efficient surface cleaning can be accomplished by dissolving the oil or grease (water-soluble contaminant) from the surface of the workpiece. Then the metallic residue (water insoluble contaminant) will detach from the part since the sticky medium is no longer present.

Component part cleaning can be accomplished manually or through the use of specialized cleaning equipment. Manual cleaning is typically accomplished with cleaning solvents, such as freon, which is applied to the surface of the workpiece. This method of cleaning is not practical when the number of production units is high, or the cleanliness requirements are very stringent.

Precision cleaning is required for those instances where the production volume is large, and/or the cleanliness requirement is very stringent. Precision cleaning utilizes specialized cleaning equipment to thoroughly clean contaminants from the surface of a substrate (base material). For example, precision cleaning equipment is used to prepare surfaces prior to the application of a surface coating such as electroplating. Soils, oils, oxides, and metallic residue must be completely removed in order to provide adequate adhesion of the deposited metal. The cleanliness of the substrate will affect the appearance and functionality of the applied finish. If the surface on which the coating is to be plated is contaminated with soil, the applied coating will tend to peel or flake, which is cause for rejection.

Manufacturers of electronic components also rely on the efficiency of precision cleaning equipment in the production of printed circuit boards or PCBs. PCBs must be cleaned prior to the installation of electronic components. Contaminants on the PCB could short out the electrical circuitry and affect the performance of the entire assembly. Therefore, precision cleaning operations must be very efficient in order to minimize the number of rejections.

9.2.2   Examples of Precision Cleaning Equipment

There are many types of precision cleaning systems available on the market today. The most popular method of precision cleaning is vapor degreasing with chlorinated solvents. However, chlorinated solvents contribute to the degradation of the ozone layer, and therefore the production and use of these chemicals have been restricted worldwide. As society assigns new costs to use these solvents, other technologies previously considered too expensive have now become more cost-competitive. Previously neglected cleaning methods such as aqueous and semiaqueous cleaning systems are now being evaluated by industry. However, if it were not for the environmental restrictions, industry would still prefer to use vapor degreasing with chlorinated solvents due to the simplicity of the process.

In order to completely understand industry's reliance on chlorinated solvents, each of these cleaning methods must be explored individually. The following sections provide detailed information regarding vapor degreasers, and aqueous and semiaqueous cleaning systems.

Vapor Degreasing Properties. All vapor degreasers are very similar in design and use chlorinated solvents to remove grease, oil, and other contaminants from the surface of a workpiece. Chlorinated solvents, described earlier in Section 9.2, are used in vapor degreasing units because of the high density of their vapors, good solvency for organic materials, low latent heat of vaporization, nonflammability, noncorrosiveness, and relatively high stability.

Each of the chlorinated solvents provides good cleaning for normal industrial soils. However, each solvent has its own advantages that dictate its usefulness in a particular application. Trichloroethylene is the solvent of choice for removing paint films, heavy resins, and buffing compounds. For cleaning printed circuit boards, electronic components, and electrical motors, 1,1,1-trichloroethane and blends of trichloroethane are well suited. Because of its low boiling point, methylene chloride works very well for temperature-sensitive parts and where aggressive solvency is required. Perchloroethylene, because of its high boiling point, is excellent for removing high-melt waxes and for cleaning light-gauge metal parts.

A typical vapor degreasing unit consists of an open-top tank with heaters at the bottom and cooling coils at the top of the unit. The source of the heat can be electricity, steam, gas, or superheated water. Pure vapors are generated by heating the chlorinated liquid solvent to its specified boiling point. The solvent vapors rise and displace the air in the tank. The vapors are confined by cooling coils located in the upper section on the degreasing unit. When the solvent vapors reach the cooling zone, they condense on the cooling coils. The solvent is then collected, routed through a water separator, and returned to the boiling sump of the degreaser.

A vapor degreaser is very simple to use. The work is lowered into the vapor zone of the degreaser unit, where the solvent condenses on the cool part, flushing off the oil, grease, and soil contaminants. Cleaning will continue until the temperature of the work is equivalent to the temperature of the vapors. At that point, condensation ceases, and the part is lifted to a freeboard region in the tank in order to dry the workpiece. If additional cleaning is required, the work is lowered into the boiling solvent, or sprayed with solvent at a high pressure. Once the part is clean and dry, it is removed from the degreasing unit and sent to the next workstation. The entire cleaning and drying operation is accomplished on one machine in a matter of minutes.

The freeboard zone, measured from the top of the vapor zone to the top of the tank, permits the drying of the work before it is removed from the degreasing unit. The freeboard area is also very useful in minimizing vapor disturbance from air movement. Both of these factors reduce the vapor emissions. OSHA and the EPA require the freeboard height to be a minimum of 75% of the width of the tanks.

The design and operation of a degreaser makes a very good candidate for automation. In large plants, work can be positioned in the degreaser through the use of monorails, conveyors, or roller systems. The appropriate degreasing cycle can then be incorporated into the production line to permit continuous cleaning operations. However, one of the more common efforts in mechanized vapor degreasing is neglecting to maintain the work in the freeboard region of the degreaser for a sufficient amount of time. Thereby, not allowing the work to dry, and allowing the chlorinated solvent to evaporate into the environment. This can be hazardous to personnel operating the unit, as well as detrimental to the ozone layer.

Equipment. There are three basic types of vapor degreasing units available. They are: 1) vapor, 2) vapor-spray-vapor, and 3) liquid immersion-vapor. The variations are simply alterations of the cleaning cycle. For example, to clean work that is heavily caked with grease and/or soil, it may be necessary to spray the work with solvent at high pressure, or immerse it in a boiling sump.

Simple Vapor Degreasing. This type of vapor degreasing machine utilizes a straight vapor cycle. Vapor degreasing usually takes only 10 minutes or less and there is no mechanical action, except for solvent runoff. The wok is immersed in pure solvent vapors and the parts are cleaned when the vapors condense on the surface of the part, dissolving the soluble contaminants and flushing away the insoluble. Not all insolubles, such as rust and mill scale, can be removed by vapor degreasing. Furthermore, it is necessary for the parts to be arranged separately so that condensation from one part to another does not drain onto each other. The amount of condensation depends on the solvent used and the weight of the specific heat of the work load being cleaned. Vapor-type degreasers are effective when there is a thin deposit of soluble soil.

Vapor-Spray-Vapor Degreasing (VSV). The vapor-spray-vapor technique assists in the mechanical removal of insoluble soil. Chlorinated solvent is pressure-sprayed on the work to remove materials that are adhered to the workpiece. Typically the vapor spray is controlled manually with a spray lance. Conveyorized degreasers use spray nozzles, which are directed at the work as it passes through the degreaser. The spray also reduces the temperature of the workload prior to the final rinse. The size of the workload and utilization of the degreaser are limited by this cleaning cycle. The VSV cycle does not lend itself to cleaning of closely nested work such as screw machine parts.

Immersion-Vapor Degreasing. The immersion-vapor technique provides more mechanical action in order to clean heavily contaminated workloads. This technique permits rapid solution of contaminants and removal of insoluble contaminants such as chips, excess buffing, and grime by means of mechanical action of the boiling solvent. This method is used when the workload can be handled in baskets or bulk containers. The workload is immersed in the sump with the boiling solvent, and then immersed in a cool sump. The up-and-down motion tends to remove insoluble contaminants. The cool sump reduces the temperature of the workload prior to the final vapor rinse and drying. This method may be impractical for large workpieces.

9.2.3 Alternatives to Chlorinated Solvent Degreasing

In selecting an alternative degreasing technology, one should remember that vapor degreasing was so popular because it was inexpensive and simple. As society assigns new costs to these solvents, other technologies previously considered too expensive have now become more cost competitive. Change will continue to flourish in the foreseeable future. As previously neglected solvents are explored, they will be closely examined for health, safety, and environmental impact. —W. P. Innes [7]

There are basically two types of alternatives to chlorinated solvents currently available for precision cleaning operations: alternative cleaning processes and alternative cleaning solvents. A compilation of the more popular alternatives is presented here [5]:

1. Alternative cleaning processes:
- Aqueous
- Semiaqueous
- Pressurized gasses
- Supercritical fluids
- Plasma cleaning
- Ultraviolet/ozone
2. Alternative cleaning solvents:
- Hydrochlorofluorocarbons (HCFCs)
- Alcohols with perfluorocarbons
- N-Methyl-2-pyrrolidone
- Aliphatic hydrocarbons
- Miscellaneous (i.e., ketone and ethers)

The substrate that is to be cleaned, type of contaminant, and the utilization of the cleaning system must all be considered when evaluating which of these alternatives is the best. However, alternative cleaning processes such as aqueous and semiaqueous cleaning are quickly becoming industry's standard. Cleaning processes such as pressurized gases, supercritical fluids, plasma cleaning, and ultraviolet/ozone methods have not gained acceptance throughout industry due to their ineffectiveness, high capital costs, uneven cleaning rates, and size. The alternative solvents are also unattractive to the cleaning consumer due to the fact that the production and use of many of the replacement chemicals have been restricted due to environmental regulations.

9.2.4 Aqueous Cleaning

Aqueous cleaning system (ACSs) combine non-ozone-depleting solvents (detergents) and surfactants in an aqueous base. The aqueous process emulsifies and solubilizes grease and oils as well as the chlorinated solvents previously discussed. However, the cleaning process is much more complex and involves significant costs associated with waste treatment of the contaminated water solution. The advantages and disadvantages of aqueous cleaning systems are discussed next.

The key stages of an aqueous cleaning process include washing, rinsing, drying, and wastewater disposal. A separate piece of equipment is required for each of these processes. Although each step is an important and integral part of an aqueous cleaning system, rinsing and drying may not be necessary for all applications. Furthermore, wastewater disposal may be reduced by recycling the bath contents through the system several times.

The aqueous cleaning tank, rinse tank, and dryer are positioned in-line provided that there is sufficient space on the shop floor. This is because the work that is to be cleaned must proceed through each of the tanks, in order. There is no back flow. However, if the cleaning system, consisting of each of the aforementioned tanks, can be characterized as high utilization, then it may be more efficient to integrate it within the production line. If the utilization is very low, then the aqueous cleaning system will occupy its own individual cell; this is referred to as batch type.

In-line and batch-type equipment can be further characterized into *immersion, spray,* and *ultrasonic* type equipment. The advantages and disadvantages are discussed.

- *Immersion* equipment cleans parts by immersing them into a solution and using some form of agitation to displace and float away contaminants. Soil is removed from a metal surface by convection currents in the solution, which are created by heating coils or mechanical action.
- *Spray* equipment washes parts with a solution sprayed at high pressures. The important parameter is the kinetic energy imparted to the cleaning fluid at the point of impact. Spray pressure can vary from 2 psi to 2000 psi and more. Obviously, the higher the pressure, the more mechanical help is provided in the removal of soil from a metal surface.
- *Ultrasonic* equipment works well with water-based processes. This is due to the high cavitation efficiency in aqueous solutions as compared to organic solvent media. Ultrasonic activity in aqueous solutions is typically 7–10 times higher than the activity in chlorinated solvents. It should be noted that the design of ultrasonic cleaning equipment must take into account that too much cavitation may erode metal surfaces, and that the geometry of some parts is more sensitive to ultrasonic processes [8].

Advantages of Aqueous Cleaning Systems.

- *Safety*—Aqueous systems have few problems with worker safety as compared to other ones. They are not flammable or explosive. It is important, however, to consult the material safety data sheets for information on health and safety.

- *Cleaning*—Aqueous systems can be readily designed to clean particles and films better than solvents.
- *Multiple Degrees of Freedom*—Aqueous systems have multiple degrees of freedom in process design formulation and concentration. Thus, aqueous processes provide superior cleaning for a wider variety of contamination.
- *Inorganic or Polar Soils*—Aqueous cleaning is particularly good for cleaning inorganic or polar materials. For environmental and other reasons, many machine shops are using or converting to water-based lubricants and coolants versus oil-based ones. These are ideally suited to aqueous chemistry.
- *Oil and Grease Removal*—Organic films, oils, and greases can be removed very effectively by aqueous chemistry.
- *Multiple Cleaning Mechanism*—Aqueous cleaning functions by several mechanisms rather than just one (solvency), including displacement, emulsification, dispersion, and others. Particles are effectively removed by surface activity coupled with the application of energy.
- *Chemical Cost*—Low consumption and inexpensive.

Disadvantages of Aqueous Cleaning Systems.

- *Cleaning Difficulty*—Parts with blind holes and small crevices may be difficult to clean and may require process optimization.
- *Process Control*—Aqueous processes require careful engineering and control.
- *Rinsing*—Some aqueous cleaner residues can be difficult to rinse from surfaces. Nonionic surfactants are especially difficult to rinse. Trace residues may not be appropriate for some applications and materials. Special precautions should be applied for parts requiring subsequent vacuum deposition, liquid oxygen contact, etc. Rinsing can be improved using deionized (DI) water or an alcohol rinse.
- *Floor Space*—In some instances, aqueous cleaning may require more floor space.
- *Drying*—For certain part geometries with crevices and blind holes drying may be difficult to accomplish. An addition drying section may be required.
- *Material Compatibility*—Corrosion of metals or delayed environmental stress cracking of certain polymers may occur.
- *Water*—In some applications high-purity water is needed. Depending on purity and volume, high-purity water can be expensive.
- *Energy Consumption*—Energy consumption may be higher than solvent cleaning in applications that require a heated rinse and drying stages.
- *Wastewater Disposal*—In some instances use of aqueous cleaning may require wastewater treatment prior to discharge.

9.2.5   Semiaqueous

Semiaqueous cleaning systems are very similar to aqueous cleaning systems. The only difference is the type of solvent used to perform the cleaning. Semiaqueous systems use hydrocarbon/surfactant cleaners in lieu of detergents or chlorinated solvents for precision cleaning applications.

Hydrocarbon/surfactant cleaners are used in cleaning processes in two ways. They are either emulsified in water solutions and applied in a method similar to the aqueous cleaners, or they are applied in concentrated forms and then rinsed with water. Since both methods use water in the cleaning process, the hydrocarbon/surfactant process is referred to as a semiaqueous process.

The four major steps in semiaqueous cleaning are similar to those mentioned for the aqueous cleaning process. First the workload is washed with a hydrocarbon/surfactant cleaner, rinsed with water, dried, and finally the wastewater is treated. A separate piece of equipment is required for each process. The advantages and disadvantages of the semiaqueous cleaning process will be discussed.

As with aqueous cleaning, the semiaqueous cleaning equipment can be set up as an individual cell or within the production line. Furthermore, it can be classified as either spray or immersion equipment. However, it should be noted that it is potentially dangerous to use existing aqueous cleaning equipment with semiaqueous cleaning solutions due to the potential for flammability.

Mechanical spray action can improve the performance of the semiaqueous cleaning process. However, when using the concentrated hydrocarbon/surfactants, the atomized solution will tend to combust and special care must be taken to preclude the occurrence of a fire. Nitrogen blanketing can be used to remove oxygen from the spray chamber to reduce the possibility of a fire.

Immersion equipment is the simplest design for semiaqueous cleaning, and it does not present the combustion danger of concentrated hydrocarbons/surfactants. This is due to the fact that immersion equipment uses dilute emulsion solutions. Some old vapor degreasers may be retrofitted to immerse the parts into a bath of emulsion cleaner. The parts would simply be cold dipped into a bath, which may or may not be heated. Very little mechanical energy would be required to achieve cleanliness due to the solvency of the hydrocarbons/surfactants. However, higher levels of cleaning can be attained with the addition of agitation induced either mechanically or with ultrasonics [8].

Advantages of Semiaqueous Cleaning.

- Good cleaning ability (especially for heavy grease, tar, waxes, and hard-to-remove soils).
- Compatibility with most metals and plastics.
- Reduced evaporative loss.
- Decreased solvent purchase cost.
- Nonalkalinity of the process, preventing etching of metals.

Disadvantages of Semiaqueous Cleaning.

- Recycling or disposal of wastewater makes the process less economically viable.
- Some cleaners are volatile organic compounds (VOCs), which have environmental restrictions.
- Surfactants are difficult to remove.
- Drying equipment may be required.
- Some cleaners can gel in low-water solutions.

9.2.6   Problems Associated with Aqueous Cleaning Systems

Aqueous cleaning systems can alleviate industry's dependence on chlorinated solvents: however, aqueous cleaning systems have not gained industrial acceptance.   The reluctance of industry to convert to an aqueous system is based on the fact that these cleaning systems are not as simple to use as was vapor degreasing. To clean a part with a vapor degreaser, the operator simply immersed the part in the vapors and the part was clean and dry. However, with aqueous systems, the part must be cycled through several tanks, and held in each one for a specific duration of time.

A summary of some additional concerns relating to the conversion from vapor degreasing to aqueous cleaning is as follows:

- The design of a product to be cleaned may have significant influence on its ability to be cleaned. The materials and configuration should be reviewed to determine if it is possible to make changes that will influence the success of aqueous cleaning. Care must be taken to preclude cleaning fluid from becoming trapped in holes and capillary spaces. Penetration into small spaces is a function of surface tension, viscosity, and capillary forces. A cleaning solvent with low surface tension may penetrate yet it may not be easily displaced by the higher surface tension of the pure water.
- Aqueous cleaning is more complex than chlorinated solvent cleaning. Aqueous cleaning requires engineering involvement and process control for prevalent problems.   The temperature, pH, agitation, cycle time, and cleaning bath quality are important parameters that must be monitored and maintained during aqueous cleaning.
- Drying represents one of the major challenges when cleaning complex parts in an aqueous cleaning system.   Considerable engineering time and experimentation may be required to resolve the drying dilemma. However, there is a tremendous variety of spot-free drying systems commercially available.
- Industry's main concern is that aqueous cleaning systems require substantial wastewater treatment. When vapor degreaser solvents become contaminated, the degreasing unit is drained, cleaned, and refilled with clean solvent. The used solvent is packaged and hauled away as hazardous waste. Sure, it is costly, but it was simple and there was significantly less contaminated fluid to dispose of in comparison to the aqueous processes. With the aqueous

cleaning process, the company must allocate capital and floor space to support the treatment of the wastewater. This is the most significant impact to the company since the amount of water used in the processes is substantial, and the cost to treat the wastewater is approximately half the original cost of the cleaning solutions [8].

9.2.7 Problems Associated with Semiaqueous Cleaning Systems

Semiaqueous cleaning methods have not gained industrial acceptance as well as aqueous systems. Semiaqueous solvents require much more wastewater treatment, which makes the process less economical.

However, the most substantial drawback is that some of the cleaning solvents are volatile organic compounds (VOCs). The EPA has established restrictions on the amount of VOCs that a company can emit to the environment. Corporations looking to invest in replacement systems for a vapor degreaser are reluctant to select an alternative that is also controlled by environmental regulations [8].

9.3 ENVIRONMENTAL CLEANING SYSTEM METHODOLOGY

Conversion from a vapor degreaser to an aqueous cleaning system (ACS) to precision-clean components must follow a few basic steps. The first step is to determine the ideal temperature, pH, and agitation of the ACS solvent for the parts that are to be cleaned. In most cases the manufacturer of an ACS can provide suggestions for these parameter settings; however, experimentation may be required to isolate the optimum parameters to precision-clean specialized materials.

9.3.1 Process Control in ACS

Once the parameter settings have been optimized, then control of the ACS processes must be established. Aqueous cleaning is much more complex than vapor degreasing from the standpoint of process control. With a standard vapor degreaser, the operator would simply turn on the power that would energize the heating coils and the refrigeration unit (for condensing coils). When the machine was not in use, it was simply covered so that the chlorinated solvents would not evaporate.

However, to efficiently clean parts with an ACS, it is essential to control the solvent temperature, pH, agitation, and rinse- water quality. This can be accomplished through the use of computer-integrated manufacturing (CIM). The parameter settings can be monitored and controlled by a computer dedicated to the ACS. The computer could be programmed to monitor the condition of the fluid in the cleaning tanks and make corrections by adding chemicals, turning the heating coils on or off, or increasing the agitation. The computer could also control the cycle time through the ACS tanks if the system was automated. This process is similar to the process by which a central processing unit (CPU) on a computer numerical control machine orientates the table and the cutting tool for machining operations.

9.3.2   Wastewater Treatment

Oil and grease are the most common contaminants that have to be removed from the surface of a workpiece. Thus, the wastewater generated from the ACS will contain a substantial amount of oil, which needs to be treated prior to the disposal of the water. There are four methods currently used to treat wastewater contaminated with oil and grease. These methods are:

- Gravity separators
- Ultrafiltration
- Coalescing
- Chemical treatment

Gravity separators are the most common method used by industry to treat wastewater contaminated with oil. The process takes advantage of the difference in specific gravity between water and oil. The treatment generally involves retaining the oil-contaminated wastewater in a holding tank and allowing gravity separation of the oil and water to naturally take place. The oil is then skimmed from the top of the wastewater in the tank. The waste oil is then discarded as hazardous waste.

Ultrafiltration is a low-pressure filtration process for separating high-molecular-weight dissolved materials from liquids. A semipermeable membrane performs the separation. The wastewater is pumped tangentially through tiny holes in the membrane. However, oils and suspended solids become trapped. The waste oil is then discarded as hazardous waste.

Coalescing involves the preferential wetting of a coalescing medium by oil droplets that accumulate on the medium. These droplets rise to the surface of the solution as they combine with other, larger particles. The oil is then collected and discarded as hazardous waste.

Chemical treatment is often used to breakdown wastewater that is contaminated with oil and grease. The primary disadvantage of this process is that the use of these chemicals can generate a sludge that must be disposed of a hazardous waste. It is also the most expensive process of those just described.

All of the aforementioned waste treatment processes involve the disposal of hazardous waste. This can be very costly, and it certainly not good for the environment. One such alternative, which is not currently used in industry, for ACS wastewater treatment is microbial oil remediation (MOR). MOR essentially means "bacteria that eats oil." This bacteria, referred to as a microbe, literally devours oil and other organic compounds such as grease, paraffin, and aromatic hydrocarbons. The microbes eliminate oil safely, quietly, and without creating environmental hazards. The micro-organisms convert these contaminants into harmless, naturally occurring products that are environmentally safe. The process is similar to composting a pile of leaves.

9.4   ENVIRONMENTALLY SAFE DISPOSAL OF AEROSOL CANS

The 1990 amendments to the Clean Air Act have significantly reduced the amount of volatile organic compounds that can be contained in products, and disposed of by industry. These new regulations will affect the products packaged in aerosol spray cans since they typically have a high VOC content.

Household products such as hair spray, deodorant, and oven cleaners, as well as industrial products such as paint, lubricants, and cleaners, all contain a high VOC content. These products are regularly packed in aerosol cans due to the desire for convenience. However, the disposal of aerosol cans will become more difficult with the passage of more stringent environmental regulations.

The U.S. Environmental Protection Agency (EPA) specifies that aerosol cans will be regulated as hazardous waste under the Resource Conservation and Recovery Act (RCAA) (only if such cans exhibit a hazardous waste characteristic). Further, the EPA indicates that there is no data to indicate that spent aerosol containers exhibit a hazardous waste problem. On May 19, 1992, California adopted regulations (Section 66261.7) that would exempt emptied aerosol cans from regulation as hazardous waste. Therefore, hazardous waste control laws do not apply to aerosol cans that are emptied through normal use. This regulation set a standard for the rest of the nation to follow.

Aerosol cans that could not be emptied through normal use, possibly due to a defective nozzle, would have to be managed as hazardous waste. This can result in significant expenses for a company that relies on products packaged in aerosol spray cans to assist in production.

RCAA regulations require that, unless relieved of internal pressure, aerosol cans must be placed in a drum and manifested for solid hazardous waste disposal. A typical 55-gallon drum can hold 96 aerosol cans and would cost as much as $1500 for proper transportation and disposal. This means that the typical aerosol paint could cost as much as two to three times more to dispose of than their original purchase price.

As realized, an alternative method of disposing of aerosol cans must be developed to satisfy the needs of industry worldwide. The alternative method should eliminate the tremendous expense associated with the disposal of aerosol spray cans without having a detrimental effect on the environment.

9.4.1   Why Do We Need Aerosol Spray Products?

Many products are packaged in aerosol spray containers for convenience. Alternate methods of packaging have been introduced into the market, such as "the pump," and have been successful for some applications. For example, hair spray and deodorant manufacturers have converted to plastic containers with the push-button pump to eliminate their dependence on aerosol cans. However, these alternatives actually contain more VOCs than their aerosol counterparts because they use alcohol as a drying agent and a solvent. The companies that have converted to the pump-type packaging have done it because it makes their product more appealing to the "environmentally correct" consumer.

Furthermore, for paint products, the conversion to alternative packaging is not as easy. Aerosol spray paints rely on the propellant to act as a lubricant to keep the paint from clogging the ejection port. Therefore, replacing the aerosol can with an alternative packaging method would have to be done on a case basis since the performance of some products relies on the aerosol container.

9.4.2   Perceptions of the "Environmentally Correct" Consumer May Be Wrong

"Environmentally correct" consumers who think that aerosols are the leading threat to the ozone layer are mistaken. The use of 98% of the ozone-depleting CFCs in aerosol cans was banned in Norway, Canada, Sweden, and the United States in 1978. Nevertheless, spray cans still have a bad reputation [10].

According to recent surveys, most Americans incorrectly believe that aerosol cans are dangerous to the environment. The aerosol industry claims that the myth is perpetuated by the media, who seem to be unaware that aerosols have not contained CFCs for 16 years. Furthermore, a 1991 survey revealed that Americans were unaware that refrigerants used in car air conditioners, solvents, filler material in foam products, and cleaning agents release CFCs and pose a far greater threat to the ozone layer than do aerosols. Many alternatives to CFCs are available, but companies are not committed to make the switch [11].

As seen from the information just presented, the perception of the environmental hazard associated with aerosol products may be exaggerated by the media. The environmental hazard associated with aerosol products is twofold:

1. Most contain volatile organic compounds (VOCs).
2. Partially filled cans may contain propellant, making the cans reactive if put in contact with a strong igniting force such as heat or intense pressure.

9.4.3   What Has Been Done to Reduce VOC Content in Aerosol Cans?

Environmental regulations are gradually reducing the percentage of VOCs in aerosol products. The California Air Resources Board (CARB) limits the amount of VOCs in all consumer products. The permitted level is 80% as of January 1, 1993, 55% as of 1995, and 0% by the year 2000. These standards present a challenge for spray manufacturers who produce products that maintain performance with much lower ethanol and hydrocarbon contents [12].

Further, the 1990 amendments to the Clean Air Act reduce the VOC content of aerosol spray paint from 5.0 lbs/gallon within 2–3 years. This reduction in VOC content may adversely affect the aerosols paint performance since the paint manufacturer will have to increase the percentage of solids in oil-based paints, which will result in the paint being thicker and much harder to apply [13].

9.4.4   Beating the Regulations

The current environmental regulations specify that if an aerosol can is empty, (i.e., with an internal pressure equal to atmospheric pressure) then the aerosol can may be disposed of as nonhazardous waste. The rationale for this regulation is that the small amount of VOC contents remaining in the depressurized can is inconsequential (less than 3% by weight is considered to be nonregulated by the EPA). Further, if the can is at atmospheric pressure, then it does not pose the threat of igniting if subjected to a strong igniting force, such as high heat.

Several companies have taken advantage of the idea and developed products that would provide assurance that the aerosol can will in fact be empty and depressurized after processing through the use of their equipment. Basically the device will puncture the can and drain its contents. This will relieve the pressure in the can while its contents can be collected. The following sections describe several of these devices.

Can Emitter. American Gas Products (Whittier, California) manufactures a device called the can emitter. This device will function exactly as previously described. The unit mainly consists of a two-part cylindrical chamber with a hand-operated punch on the side of the device. The unit is quite simple to operate. Simply insert an aerosol can within its lower chamber, lower a cover over the can, and turn a threaded punch into the can. The aerosol can will depressurize with 5–10 seconds, and the contents will drain out. The waste drained from the cans will be collected in a hazardous waste container; and the emptied aerosol can may be disposed of as nonregulated waste or it may be recycled.

Although this unit seems to be quite simple, it effectively meets the current EPA regulations. However, the can emitter unit does not preclude the potentially ignitable fumes from within the aerosol cans from becoming a hazard, nor does it capture all of the VOCs that were contained in the can.

Aerosolv Aerosol Can Disposal System. Justrite Manufacturing Company (Des Plains, Illinois) has developed a product called Aerosolv that punctures the aerosol can, collects the hazardous contents, and filters the airborne organic compounds.

The Aerosolv unit is easily installed and is operated by hand. It is made of aircraft aluminum for durability. It consists of two parts: a puncturing unit and a filter. The puncturing unit will thread directly into a 2-inch bung of a 55-gallon drum, and the filter directly threads into a ¾-inch bung on the opposite side of the drum.

Aerosolv is easy to operate. The aerosol can is inserted into the puncturing unit (inverted), a protective cover is lowered over the can to secure it in place, and a hand-operated lever is used in the puncturing operation. It will take between 5 and 10 seconds for the can to be depleted after puncturing. Once the process is complete, the spent aerosol can remains in the puncturing unit until another can is ready to be processed. This precludes the possibility of fumes or VOCs escaping from the drum without being filtered. The drum cannot become overpressurized, since the filter is designed to relieve the internal pressure at 3 psi.

The most significant advantage that Aerosolv has over the can emitter is the filter. The Aerosolv filter was designed specifically for removing ignitable vapors from the pressurized gas prior to emission through the ¾-inch bung of the drum. The filter is composed of two parts: a coalescing lower level and an activated carbon upper level.

EVAC Aerosol Can Crusher and Recovery System. Beacon Engineering Company (Jasper, Georgia) has developed a more elaborate system for dealing with spent aerosol cans. There are basically two systems available from Beacon: EVAC Jr., which processes 200–300 cans per hour, and EVAC II, which processes 1200–2000 cans per hour.

The Beacon EVAC system is the only system that captures (under pressure), encloses, and retains up to 98% of all the contents from the disposed aerosol can. The EVAC system was developed to satisfy the concerns regarding the release of harmful VOCs into the atmosphere as well as plant and operator safety.

The EVAC can crushers will automatically feed and crush cans in multiples or singles, into ½-inch thick wafers. The concentrate can be reclaimed or disposed of like other bulk waste. The propellant can be separated from the product concentrate and burned off at a remote site.

9.4.5   Can Recycling

An aerosol can, when empty, is merely another steel container that should be recycled. Steel can recycling has grown in the past 3 years to a rate of 25%, with aerosol cans becoming an increasing fraction of that percentage. Given the recycling option, steel cans of any type should no longer be regulated to landfills.

Aerosol cans in the industrial shop or factory may not be empty. However, the cans could be more easily recycled or otherwise disposed of if they are not punctured, drained, and flattened with a crushing unit. This allows their accumulation together with other ferrous scrap routinely generated in the plant and sent to the local scrap dealer. Alternatively, the cans be accumulated separately for transportation to a waste handler. Further, aerosol cans in the home and small businesses that are emptied through normal use could be magnetically separated along with steel food and beverage containers from other materials, such as glass, aluminum, and plastic.

9.5   APPLICATION OF PAINTING TECHNIQUES THAT REDUCE VOC EMISSIONS

In the past two decades, antipollution laws have mandated important changes in paints and coatings, notably the removal of lead and mercury. Today, a host of new environmental regulations is forcing yet more changes. These regulations concern volatile organic compounds (VOCs). VOCs are emitted into the atmosphere as paint dries and reacts with other chemical compounds and ultraviolet radiation to form ground-level ozone, commonly called smog [2,3,14,15].

To comply with clean air mandates, California, New Jersey, and a number of other major metropolitan areas have already set limits on VOCs in paint. Many other jurisdictions are considering measures. The problem of paint emissions illustrates the seemingly endless intricacy of environmental pollution and our efforts to control it. Paint manufacturers, while reluctant to seem antienvironmental, complain that their industry is being overburdened as compensation for our failure to regulate automobiles. Paint manufacturers argue that the result of these regulations may be to replace paints of proven durability with inferior products that need frequent painting. Since VOCs are emitted when newly applied paint dries, this might mean more, not less, pollution. VOCs in factory-applied coatings have been regulated for years, but throughout the United States, these limits are becoming stricter.

Paint consists essentially of a pigment, a vehicle or thinner, and a resinous binder. Despite their variety, all paints can be broadly categorized by their vehicle, as either solvent-thinned or water-thinned. VOC regulations affect primarily solvent-thinned paints. Because of the EPA's deadline for issuing a national standard, it will be several years before the full effect of VOC regulations are clear.

9.5.1   Conventional Solvent-Based Processes and Emission Points

Some common methods of applying coatings to miscellaneous parts are described here. Each of these applications uses solvent-based paints and emits unwanted amounts of VOCs into the atmosphere.

- To apply a flow coating, the metal parts are moved by a conveyor through an enclosed booth. Inside, nozzles (which may be stationary or may oscillate), located at various angles to the conveyor, shoot out streams of paint that flow over the part. The excess paint drains into a sink at the bottom of the booth, is filtered, and is pumped back for reuse. This method provides about a 90% transfer efficiency. The coated parts are often conveyed through a flashoff tunnel to evaporate solvents and allow the coating to flow out properly.

- To apply a dip coat, the metal parts are briefly immersed either manually or by conveyor into a tank of paint. The excess is allowed to drip from the part and drain back into the tank. This method provides about a 90% transfer efficiency of the coating. The viscosity in dip coating is very critical.

- Spray painting is the most common technique for applying a single coat, some primers, and most topcoat applications. Its transfer efficiency is 40–70%. Electrostatic spraying with disc, bell, and other types of spray equipment are commonly used to increase transfer efficiency to 90%. Transfer efficiencies vary with parts being coated and the expertise of the operator. After painting and flashoff, the parts are often baked in a single or multipass baking oven at 275–450°F.

9.5.2   Applicable Systems of Emission Reduction

Some common methods of applying coatings to miscellaneous parts are described next. Each of these applications uses reduced solvent-based or non-solvent-based paints and emits little or no VOCs into the atmosphere.

- Waterborne or perhaps more truthfully "water-thinned" paints are paints that use water to reduce the solvent content in the coating. However, some volatile solvent is still present, perhaps as much as 10% by volume.

- Waterborne paint (spray, dip, or flow coat) can be used in an oven-baked single coat, primer, and topcoat, and air-dried primer with a possible reduction in VOCs of 60–90%, depending on the composition of coating, transfer efficiency, and film thickness.

- Waterborne electrodeposition is used in an oven-baked single coat and primer with a possible reduction in VOCs of 90–95%, depending on the composition of coating, transfer efficiency, and film thickness.
- Higher solids are paints with increased solid contents. By increasing the solid content, the other ingredients that make up the paint, like solvents, are decreased.
- Higher solids (spray) is used in an oven-baked single coat and topcoat, and air-dried primer and topcoat with a possible reduction in VOCs of 50–80%, depending on the composition of coating, transfer efficiency, and film thickness. Also, figure in the expertise of the operator applying the coating.
- Carbon absorption (add-on equipment) is used with an oven-baked single coat, primer, and topcoat application and flashoff areas as well as air-dried primer and topcoat application and drying areas. This reduction in VOC emissions is 90% only across the control device and does not take into account the capture efficiency. This requires extensive duct and air-handling equipment.
- Powder coatings are used for an oven-baked single coat and topcoats with a possible reduction in VOCs of 95–98%, depending on the composition of coating and film thickness.

9.5.3 VOC Regulations in Spray Painting

In the area of regulations on VOC emissions from industrial coating operations, three types of rules are presented:

- Rules restricting VOC emissions from coating operations in steel fabricating and painting shops. This is typically referred to as VOC emissions from surface coating of miscellaneous metal parts and products.
- Rules restricting VOC emissions from coatings intended to be applied in the field to bridges and industrial structures. This is typically designated as a subcategory or series or subcategories with architectural coatings rules.
- Rules restricting VOC emissions applied in shipyards to marine vessels. This is typically designated as marine coatings rules.

9.5.4 Rules on Surface Coating of Miscellaneous Metal Parts

Of the three types of current rules, the most common type is the miscellaneous metal parts and products rules, which apply to coating operations in fabricating and painting shops and usually include rail car shops and other steel fabricating facilities.

Typically, these rules are based on an amendment to the 1990 Clean Air Act that regulates VOC emissions called control technique guidelines (CTG). Coatings are classified as follows, with their VOC limitations, based on pounds of VOC per gallon:

- Air-dried coatings: 3.5 lbs./gallon (420 g/L), dried by the use of air or forced warm air at temperatures up to 194°F (90°C).

- Clear coatings: 4.3 lbs./gallon (520 g/L), unpigmented or transparent coating, lacing color and opacity.
- Extreme performance coatings: 3.5 lbs./gallon (420 g/L), designed for harsh exposure or extreme environmental conditions.
- All other coatings: 3.0 lbs./Gallon (360 g/L), any other type of coating.

9.5.5   VOC Test Methods for Surface Coatings

The regulation of the amount of VOCs that can be released into the atmosphere has seen the development of test methods that allow manufacturers, users, and regulators of coatings to determine whether or not products comply with these regulations. These test methods have been developed to meet the changing needs of our industry but will need to be continually updated to keep pace with changing regulations and coating technologies. VOC content of surface coatings can be tested with any of several methods, depending on the regulations to be met and/or the generic type of coating. Four of the most widely used methods for determining the VOCs of coatings are listed as:

- ASTM Method D 3960-89, "Practice for Determining Volatile Organic Compound (VOC) Content of Paints and Related Coatings."
- Environmental Protection Agency (EPA) Reference Method 24.
- South Coast Air Quality Management District (SCAQMD) Rule 107.
- Bay Area Quality Management District (BAAQMD) Methods 21 and 22.

9.5.6   Why Paint Anyway?

With all of the environmental concerns and regulations that result from the application of paints, one must ask the question, Why paint it anyway? It seems that the answer to this question is that in the interest of safe plant operation, you cannot allow corrosion of steel structures to go unchecked. Therefore, a comparison of the cost of painting to the cost of replacing steel cannot really be made. Costs can be determined by using mathematical models involving the principles of engineering economy, provided that the alternative can be predicted to a reasonable degree. We must be in a position that the protective coating system must be maintained of such structures to prevent environmental corrosion that might cause failure of such structure. The consequences of potential structural problems due to inadequate maintenance could result in unsafe working conditions, interruption of plant utilities, hazardous material leaks, and violation of applicable local, state, and federal regulations. Since these consequences may have a severe economic impact, a program of painting is the optimum alternative.

9.6   SUMMARY

Industry is currently going through a serious phase of changing practice in cleaning, degreasing, and improving all manufacturing processes to be environmentally friendly.

Waste and harsh chemicals should be minimized to meet new local and state regulations. Some directions that industry has taken to minimize pollution were by changing the chemicals currently being used and by changing the manufacturing process.

From changing the use of certain chemicals to approve alternative solutes, the EPA standards are usually instantly met. The EPA periodically posts tests and studies on potential chemicals that can replace existing chemicals. With this information the company can discuss whether such a switch in chemicals is affordable, desirable, and/or desirable with their current process. For example, the aqueous cleaning systems discussed in this chapter would enable industry to precision-clean parts just as effectively as vapor degreasing with chlorinated solvents. However, the ACS would not have any detrimental effects on the environment.

The question is, Is it too late? The answer is that restrictions on chlorinated fluorocarbons (CFCs) must be implemented soon and must be observed by the whole world. The world cannot continue to be so careless with the irreplaceable resources that mother nature has bestowed upon us. Therefore, the importance of environmentally conscious engineering and design for the environment cannot be overemphasized.

9.7   BIBLIOGRAPHY

1.   Zavodjancik, J. Aerospace Manufacturer's Program Focuses on Replacing Vapor Degreasers, *Plating & Surface Finishing*, American Electroplaters and Surface Finishers Society, Inc., 79(4), pp. 26–28, April 1992.

2.   VOC Reg Update, CTGs and NESHAPs on the Horizon, Industrial Finishing Staff in Cooperation with the EPA, *Industrial Finishing*, pp. 13–31, August 1993.

3.   New Paints, Equipment Lower VOCs and Improve Product, Environmental Mandates Had Silver Lining, *Waste Lessons*, United Technologies Corporation, 2(4), pp. 1–4, Winter 1994.

4.   Composites Unit Captures Disappearing Dollars, *Waste Lessons*, United Technologies Corporation, 2(1), pp. 1–4, Winter 1994.

5.   "Solvent Cleaning (Degreasing)," Center for Emissions Control, 1992.

6.   Johnson, J.C. Chemical Surface Preparation—Vapor Degreasing, *Metal Finishing Handbook*, 1993.

7.   Innes, W.P., Metal Cleaning, *Metal Finishing Handbook*, 1993.

8.   "Eliminating CFC-113 and Methyl Chloroform in Precision Cleaning Applications," U.S. EPA, 1991.

9.   Miller, "Living in the Environment," 1990.

10.   Lefton, T. Still Battling the Ozone Stigma, *Adweek's Marketing Week*, March 16, 1992.

11.   Kaplan, J. Are the Ninja Turtles Misinformed?, *Omin*, June 1993.

12.   New Regulations Make Waves for the Hair Spray Business, *Chemical Week*, February 3, 1993.

13.   *Journal of Protective Coatings and Liners*, 8(5), February 1991.

14.   Bakke, T. Low Pollution Paints, *Home Mechanix*, March 1993.

15.   Starr, D. How to Protect the Ozone Layer, *National Wildlife*, January 1988.

9.8   PROBLEMS / CASE STUDIES

9.1   One of the main goals of the EPA is to minimize waste. EPA defines waste minimization as consisting of source reduction and recycling, with source reduction as the preferred choice. The easiest way to reduce waste, in the case of degreasing, is to minimize the use of all volatile organic compounds and other toxic cleaners. Today five types of cleaners rank at the top and are used by most manufacturing companies. These are solvents, Aakaline cleaners, acid cleaners, abrasives, and water. Discuss each of these cleaners and provide some of their uses and their limitations.

9.2   Alkaline aqueous solutions are the most common form of cleaning solutions. They are water-based mixtures with pH greater than 7 (most alkaline cleaners range from 10 to 14). This form of cleaner ranked high due to its ability to clean both lubricating oils and grease effectively. It can be designed to allow good separation between the water stream and the contaminant, oil. This allows control of the type of waste stream produced. Some of the alkaline solutions are Modern Chemical Blue Gold Industrial Cleaner, Dow Invert 1000, Turco Sprayeze LT, Borothene, NMP (*N*-methylpyrrolidone), and Daraclean 283. Discuss these chemicals, their use and limitations, and their properties such as boiling point, freezing point, flash point (flammability effect), autoignition, solubility in water, specific gravity, stability, and approximate cost per gallon.

9.3   Companies are moving from using conventional vapor degreasers to new degreasing methods due to the pressures from environmental organizations and federal agencies. These new degreasing technologies that exist and are being developed would reduce the use of solvents for many cleaning and degreasing operations. Many of these new technologies might not be ideal for all companies for many reasons. For example, high initial costs and process limitations may discourage companies and affect their decision in adopting these new technologies. Some of these technologies include add-on controls to existing vapor degreasers, completely enclosed vapor cleaners, automated aqueous cleaning, aqueous power washing, and ultrasonic cleaning. Discuss these technologies and their pollution prevention benefits, operational benefits, and limitations. Also describe the applications of these technologies.

9.4   Steel manufacturers must remove the rolling grease before the metal enters the furnace. If the rolling grease is not removed, the furnace's metal casing can corrode and hazardous fumes that are emitted to the atmosphere can be created. Vapor degreasing processes with 1,1,1-trichloroethane (VOC) can do a superb job in removing the grease, but they harm the environment.

a.   Determine an environmentally friendly degreasing method that can clean steel at a faster, better, and economical rate.

b.   Could grease separation or recycling be implemented for this method?

9.5   A company that uses approximately 10,000 aerosol spray-paint cans per year is now confronted with the environmental restrictions imposed on the disposal of these cans. The company must find an alternate means to discard these cans with increased efficiency, reduced cost, and meet the environmental regulations. In addition, the company must deal with the large number of aerosol cleaning products, lubricants, adhesives, and general-purpose types of products used on a daily basis. Most of these products will be completely empty (i.e., at atmospheric pressure) prior to disposal. However, it is highly likely that some of the aerosol products (especially spray paints) may be disposed of prematurely due to a clogged delivery nozzle or just plain ignorance. If this should happen, the company could be liable to pay a substantial fine should the EPA discover improper disposal of the aerosol cans. Therefore, precautions should be taken to preclude a violation of the environmental regulations.

a.   Assume the company is faced with the following three choices:

(i)   Do nothing, and risk paying an exorbitant fine,

(ii)   Subcontract a hazardous waste disposal firm with a premium price to dispose of the cans. At $1500 a 55-gallon drum with only 96 cans/drum, this will cost $15.63/can, or $156,300 annually based on 10,000 cans per year.

(iii)   Select one of the methodologies discussed in this chapter.

As a leader of the team evaluating these decision criteria, what would be your course of action? Why? You may wish to use the net present method of economic analysis to compare between these alternatives based on 5 years and an annual interest rate of 9%.

b.   Can additional cost savings be realized in your answer to part a? For example, can the concentrate in the 55-gallon drum be reclaimed? Can the aerosol cans be recycled? And so on.

9.6   New terpene-based solvents are claimed to provide the same functionality as harmful chlorinated hydrocarbons. Terpenes are organic solvents that are usually derived from natural sources such as pine trees or citrus fruit. Since they are made from natural sources they are considered biodegradable. Are manufacturers realizing the successes that are claimed by vendors of these products? Discuss the different types of terpenes along with their advantages and disadvantages and the areas of their applications.

TEN

DESIGN FOR QUALITY
APPLICATION IN MACHINING

What a joy it is to find just the right word for the right occasion!—Proverbs 15: 23

In this chapter, some concepts of environmentally friendly design are put into practice for the machining operations. In this way the consumer's environmental needs in machining are met by the design of a new product. The development of the concept as well as the product in terms of the quality function deployment (QFD) is presented. A practical machining problem is presented and solved through the use of QFD. The solution has led to the development of a new tool holding device. The device opens the door to other innovative, environmentally friendly manufacturing concepts.

10.1   INTRODUCTION

In the ongoing quest for continuous improvement in manufacturing operations, engineers who follow the precepts of quality engineering are recently very concerned about the environmental effect of their systems or products. There is a good reason for this concern; while features like form, fit, and function are important in defining quality for the customer, environmentally friendly products cannot be overlooked either. In fact, managers and engineers have come to realize that even products that are extremely functional cannot be successfully marketed if they have a tendency to hurt the environment.

New technologies have made our lives much easier, but past designers didn't always stop to consider undesirable side effects. The development of refrigerator compressors suitably illustrates this point: Early compressors used ammonia or sulfur dioxide, both toxic chemicals that sometimes injured and even killed people. Then chlorinated fluorocarbons (CFCs) developed and hailed for their safety and low cost were used in popular applications such as air-conditioning. Only later was CFC use connected to phenomena like global warming and ozone layer destruction.

Other areas of concern that have not been considered are machining processes, where tons and tons of cutting fluids are used daily in the United States alone. Although considerable progress has been made in ensuring the safe use of cutting fluids in manufacturing facilities, a complete solution to "eliminate" harmful effects on human life and the environment has not yet been achieved. To achieve this objective, principles of machining will be used to develop a new cutting tool holder. The new device (holder) would provide a dry, constant, and cool environment around the cutting area, inhibiting temperature rise during machining. Economic benefits include increased tool life and productivity in addition to protecting the environment from coolant waste, etc. A quality function deployment procedure (QFD) will be used to insure the quality of this device.

10.2   THE MACHINING PROCESS

Metal cutting processes are generally used to remove a thin layer of metal, the chip, by a wedge-shaped tool from a larger body, the workpiece [1]. Almost, all the mechanical energy associated with the chip formation ends up as thermal energy, causing high temperatures on the tool–work interface, and consequently affecting the tool wear rate and the machined surface. To control high temperatures, cutting fluids are used. They add many advantages such as [1–3]:

1. Improving surface finish and workpiece quality and accuracy.
2. Increasing tool life and minimizing machine down-time.
3. Reducing feed forces, cutting forces, and energy consumption and increasing machining rates.
4. Washing away the chips.
5. Protecting the newly machined surfaces from corrosion by leaving a residual film on the work surface.

Cutting fluids are classified into different groups including straight or neat cutting oils, water-based cutting fluids or soluble oils, synthetic cutting fluids, or semisynthetic cutting fluids [1,4,5]. The major actions of these groups are mainly cooling or lubrication. The cooling action involves cooling of the cutting zone, that is, the tool, the chip, and the workpiece, and thus reducing the temperatures, the tool wear, and the thermal distortion of the tool and the workpiece. The lubrication action involves reducing the weldability and the contact area between the chip and the tool surface, which results in less friction, heat, wear, and built-up edge. Also, it can modify the flow pattern around the cutting edge and increase the shear angle, which consequently reduces the generated heat at the shear plane and provides less chip thickness [1,4].

10.2.1   Functions of Cutting Fluids

To study the effect of coolants on heat removal, Trent and Smart [1,6] examined temperature contours on the rake face of high-speed steel tools used in cutting iron at 183 m/min. They conducted their experiments under three different cooling conditions: (a) dry (air), (b) wet (tool flooded with coolant over rake face), and (c) wet (jet of coolant directed at end clearance face). They found that in high-speed machining, the heat source at the chip–tool interface is a thin flow zone seized to the tool. Coolant cannot reach this zone but can only reduce the volume of the tool material, which was seriously affected by overheating by removing heat from the surrounding accessible surfaces of the chip, the workpiece, and the tool. They also found that the directed coolant jet has a better cooling effect than flooding the rake face by coolant from the op.

Trent [1] found very little effect of the coolant on the tool–work interface temperature when cooling through the workpiece or the chip. This occurs due to the high moving speed of the chip and the workpiece at the contact area without allowing enough time for heat to be transferred. Trent also noted that at high metal removal rates, all possible mechanisms of wear are very sensitive to the high temperature at the work–tool interface. For example, the rate of diffusion wear in high-speed steel tools will be doubled for an increment increase in temperature of about 20°C.

Kalpakjian [7] and Lofton [8] pointed out that at high cutting temperatures the cutting fluid is converted to a gaseous state of small molecular size that penetrates the tool–chip interface, causing more lubrication effects. However, for high-speed machining there is less probability that an effective capillary action will take place and the main role of the cutting fluid becomes only one of cooling the cutting zone.

Kurimoto and Barrow [9] studied the influence of different cutting fluids in turning of an alloy steel using a carbide tool with cutting speed range from 30 to 240 m/min. They indicated that under particular cutting conditions, cutting fluids had no lubricating action and did not penetrate into the chip–tool interface, and hence they can be considered solely as coolants. Various cutting fluids' performances were investigated on each region of tool wear: crater, flank, groove, and notch wear. It was concluded that:

- Cutting fluids can reduce the rate of crater wear.
- Dry cutting has slightly better effect than all cutting fluids on the flank wear.
- All the fluids increase the groove wear at the trailing edge, particularly at low feeds. This can be considered due to the corrosive or oxidizing action of the fluids. Since the groove wear affects the surface finish and the dimensional accuracy of the workpiece, in finishing processes tool life can decrease with the use of fluids.
- There is no significant effect of using cutting fluids on notch wear.

From these studies, it can be concluded that the main roles of the cutting fluids, under general conditions, in high-speed machining are to cool the cutting tool, decrease the crater wear, remove the chip, and cool the workpiece.

Despite these advantages and functions of the cutting fluids, they cause harmful effects to the operators and serious problems of pollution to the environment. Additional disadvantages are briefly summarized here [5,6]:

1. Potential health problems resulting from direct contact or inhalation of fluids.
2. Environmental and pollution problem resulting from degrading, recycling, disposing of unwanted waste chemicals, and treatment of waste waters.
3. Machines must be fitted with complicated systems for handling, circulating, pumping, and filtering the cutting fluids.
4. Additives must be included, such as bioresistant chemicals and rust inhibitors.
5. Damage to the tool itself in some situations. For example, in the milling process, cutting fluid may cause a big variation of the cutter teeth temperature, leading to thermal cracks.
6. In some cases, cutting fluid can cause the chip to curl into a very small radius, and thus increase the temperatures and the stresses near the tip of the tool and reduce the tool life.

To overcome these disadvantages and limitations of cutting fluids, other cooling methods were invented. The following are some of these methods.

Under cooling systems where the coolant flows through channels located under the insert then out to the environment. A thin copper foil is attached to the lower face of the insert and has two protrusions, bent downward, to remove maximum heat. The system produced larger temperature gradient and lower wear compared to the traditional upper cooling (flood cooling) method used in industry. Also, the effect of this system in thinner inserts is larger than in thicker ones when cutting at relatively high speeds [10].

Internal cooling by vaporization systems where a vaporizable liquid, such as water, is introduced inside the shank of the tool and vaporized on the underside surface of the insert. A capillary wick is used to pump the fluid automatically to this surface and there is a vent hole in the tool for the vapor generated in the cooling process. The generated heat at the cutting edge is transferred to the underside of the insert by conduction and then removed by nucleate boiling of the coolant. Although this method provides very good environmental solution, it has been shown that in heavy-duty cutting operations, the cutting temperatures are about the same as dry cutting [11].

Cryogenic systems where a stream of a cryogenic coolant, such as Freon-12 or nitrogen, is routed internally through a conduit inside the tool and directed at close range to the interface between the cutting edge and the workpiece [12].

Thermoelectric cooling systems where a module of couples of thermoelectric material elements connected electrically to one another is used [13]. When an electric current is passed through the thermoelectric elements a cold junction and a hot junction are created at the opposite ends of each of these elements. The thermocouples are so oriented that all the cold junctions are located in one side of the module, which can be called the cold face.

Cold gun air cooling systems where a vortex tube is used to cool supplied compressed air to about 30°C below its temperature without using any moving parts. The cold air is then discharged through a flexible hose and directed to the point of use. This system can be used in grinding, drilling, milling, and sawing operations [14].

High-pressure water-jet cooling systems where a high-pressure water-jet (35 to 280 MPa) is directed under the produced chip [15]. Mazurkiewicz compared this method with the dry and flood cooling methods. He pointed out in his studies [15,16] that for a given cutting speed and rake angle, this method reduces the contact length and the coefficient of friction on the tool rake face, and thus greatly reduces cutting and feed forces.

10.2.2   Goals

Based on the preceding review of the existing and the inventive cooling methods, it can be concluded that:

1. Many methods use chemical compounds or cryogenic coolants that have harmful effects on the environment and humans.
2. Some methods, such as internal cooling by vaporization and cold gun air cooling, have limited cooling effect on the cutting tools and cannot be used in heavy-duty machining.
3. Other methods, such as high-pressure water-jet cooling and internal flushing, need special techniques to be used on a large scale in industry.

Therefore, the goal of this chapter is to present a new, practical, and environmentally safe method of cooling cutting tools. The new method combines the ideas of both the internal cooling and the under cooling methods without the need for a special tool configuration but using regular tool inserts. There are many expected economic benefits from this method, such as longer tool life, increased productivity, improved accuracy, and reduced cutting fluids disposing applications. This work pertains to single-point cutting tools, such as turning tools. It focuses on orthogonal cutting processes in high-speed machining. Orthogonal cutting processes are selected for use in the present investigation because they are basically two-dimensional cutting processes. In high-speed machining, problems due to the generated heat are severe, and therefore reducing tool temperatures results in a higher metal removal rate.

10.3   TECHNICAL APPROACH

A combination of the internal cooling and the under cooling techniques already described is used in developing the new cooling methodology. Under cooling is applied to cool the lower face of the cutting insert by using a copper seat fixed to the tool holder and cooled internally so that no coolant flows out to the environment and there is no need to construct a special tiny coolant path inside the insert itself. A sketch of the new cooling device is illustrated in Figure 10.1 [32].

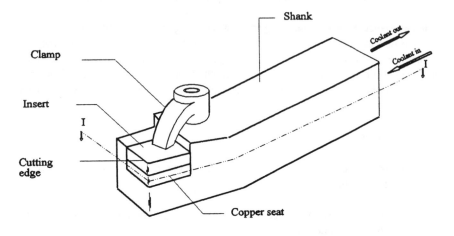

Figure 10.1. New cooling device.

The device utilizes a very small internal cooling pipe with a diameter of about 2 mm and located as near as possible to the cutting edge. A seat made of highly heat-conductive material, such as copper, is thermally connected to the lower face of the insert to increase the effective cooling area and remove maximum heat. A portion of the generated heat at the cutting zone is conducted from the tool–work interface to the insert, then into the copper seat, and consequently removed by convection using a coolant flowing in a closed cooling cycle system. The closed system uses water at room temperature to absorb heat. Water is flowing into the internal cooling pipe constructed in both of the copper seat and the tool holder. The outlet water is cooled by using a small heat exchanger and recycled again into the system with high pressure. The required pumping energy and water pressure can be estimated from the major and minor losses in the system. This energy and pressure can be varied to control the amount of heat transferred from the tool.

The device can be used in many applications such as machining of brittle material (e.g., cast iron, free machining steel, etc.), which produces discontinuous chips, and hence there is no need to use cutting fluids to wash away the chips. For continuous chips, a chip breaker can be used to decrease the chip radius of curvature and break the chips by bending [17]. The device can be tested with regular cutting inserts (high-speed steel (HSS), carbide, etc.) as well as with new cutting materials. It would provide dry and constant cooling effect on the cutting edge during machining. An air jet and/or vacuum device can be used to clear the chips from the cutting zone.

10.3.1   Thermal Analysis of the New Device

The objective of this section is to analytically determine the temperature distribution at the cutting edge in an orthogonal cutting process in three different cooling cases:

1. Dry cutting in which the tool is cooled only by natural convection.
2. Wet cutting using water-based coolant flooded directly onto the cutting tool.
3. Dry cutting using the new cooling device, which combines internal cooling and under cooling systems.

To evaluate the effectiveness of the new cooling device, a finite-element model for the orthogonal cutting process, obtained from the analytical and experimental published work of Childs et al. [18], is described next.   Interested readers could review other research in references 19, 20, and 33.

Childs' model [18] was developed for an orthogonal cutting process of an annealed medium carbon steel workpiece (0.43% C, 0.77% Mn, 0.27% Si, 0.037% S, 0.013% P, balance Fe; 165 HV50) on a lathe.   Cutting tool was an insert of Wrought BT42 high-speed steel, heat treated to 990 HV50.   Model geometry and dimensions are shown in Figure 10.2.

The model was applied to each cooling case just mentioned and a comparison between the three cases was studied.   For the case of wet cutting, cutting conditions, material properties, heat generation data, and boundary conditions were obtained from Childs et al. [18].   For the other two cases, new boundary conditions were estimated analytically to fit each case, while both of the cutting conditions and the heat generation data were assumed to be the same for simplicity.

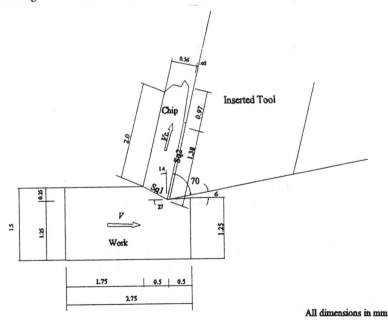

Figure 10.2. Model geometry and dimensions.

Cutting Conditions.

- Tool rake angle (α) = 14°
- Clearance angle = 6°
- Feed rate (t_o) = 0.254 mm/rev.
- Width of cut (w) = 2.54 mm
- Cutting speed (V) = 61 m/min
- Cutting force (F_c) = 1200 ± 100 N and thrust force (F_t) = 550 ± 50 N, measured with a piezoelectric dynamometer
- Friction force (F) = 870 ± 100 N, calculated from F_c and F_t
- Measured shear plane angle (ϕ) = 27 ± 2°
- Shear stress on the primary shear plane (τ_s) = 500 ± 90 MN/m², obtained from the measured forces and the shear plane angle (ϕ)
- Chip contact length = 1.38 mm, obtained from observation of the length over which material transfer occurred on the rake face
- Chip velocity (V_c) = 28 m/min, obtained from the machining test

Thermal Material Properties.

Thermal conductivity of work and chip (0.43% C steel) at 400°C = 43.6 W/m-K.

Specific heat × Density (C × p) of work and chip (0.43% C steel) = 4.3 × 10^6 J/m³-K.

Thermal conductivity of tool tip or insert (HSS) at 400°C = 22.0 W/m-K.

Thermal conductivity of tool holder (low carbon steel) at 400°C = 45.0 W/m-K.

Heat Generation Data. The localization of heat generation is approximated into the two surfaces shown in Figure 10.3 and discussed next.

The heat generation (q_1) on the primary shear plane (S_{q1}) = $\tau_s V_s$
The heat generation (q_2) on the friction plane on the tool-chip interface (S_{q2}) = $\tau_r V_c$.

where

τ_r is the friction stress distribution
τ_s is the shear flow stress of the chip, function of the applied forces and the shear angle
V_c is the sliding speed of the chip, obtained from the machining test
V_s is the velocity change from the work to the chip, and
τ_s= 500 MN/m² and V_s = 59 m/min, calculated from the machining test data

therefore,

$$q_1 = 500 \times 10^6 \times 59 \times 1/60 = 491.66 \times 10^6 \text{ W/m}^2$$

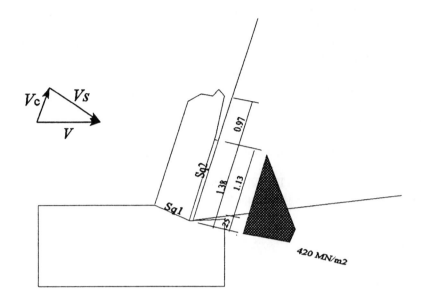

Figure 10.3. Friction stress distribution.

and the rate of total heat generated on S_{q1},

$$Q_1 = q_1 A_s = q_1 w\, t_o / \sin \phi = 698.7 \text{ W},$$

where A_s is the area of the primary shear plane.

It has been assumed that the friction stress had a uniform value along the seizure length and reduced linearly to the end of the contact length. The uniform value at the cutting edge (420 MNm^{-2}) has been obtained by matching the measured friction force ($F = 870$ N) and the friction stress integrated over the rake face. The seizure length (0.25 mm) has been assumed equal to the secondary shear zone length, obtained from the flow patterns in the quick stop section of the chip [18]. The rate of total heat generated on S_{q2} is the product of the friction force F and V_c. Thus, $Q_2 = FV_c$ = 406 W, and the rate of total energy of heat sources $Q_{total} = Q_1 + Q_2 = 1104.7$ W.

10.4   FINITE-ELEMENT ANALYSIS

The following assumptions are applied to analyze the cooling effect on the cutting edge:

- Two-dimensional steady-state conditions.
- External chip and work surfaces are assumed thermally insulated.
- Constant properties for all materials.

- Heat transfer within the tool, the copper seat, and the tool holder is by conduction only. In the moving chip and work material, heat is transferred by conduction and convection (mass transport of heat). Also, no radiation effect is considered.
- No volume heat generation within the chip or work material.
- Observed built-up-edge is included in the model and assumed to be fixed to the tool along the contact length for simplicity.
- No effect of the machine on the tool holder.

10.4.1 Governing Equation and Boundary Conditions

In each of the three regions I, II, III of Figure 10.4, the equation governing the heat transfer processes is the energy equation:

$$K \left(\frac{\partial^2 T}{\partial x^2} + \frac{\partial^2 T}{\partial y^2} \right) - \rho\, C \left(u_x \frac{\partial T}{\partial x} + u_y \frac{\partial T}{\partial y} \right) = 0 \tag{10.1}$$

where K is the thermal conductivity, T is the temperature, ρ is the mass density, C is the specific heat, and u_x and u_y are the components of the velocity in directions x and y. Their values are

$$u_x = V, \quad u_y = 0 \quad \text{in region I,}$$
$$u_x = V_c \sin \alpha, \quad u_y = V_c \cos \alpha \quad \text{in region II, and}$$
$$u_x = u_y = 0 \quad \text{in region III.}$$

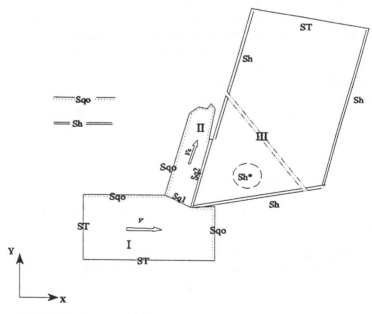

Figure 10.4. Boundary conditions.

The governing energy equation (10.1) is subject to the following boundary conditions [18]:

(i) $T = T_s$ on surfaces S_T

(ii) $-K\dfrac{\partial T}{\partial n} = 0$ on surfaces S_{qo}

(iii) $-(K\dfrac{\partial T}{\partial n})_I - (K\dfrac{\partial T}{\partial n})_{II} = q_1$ on surface S_{q1}

(iv) $-(K\dfrac{\partial T}{\partial n})_{II} - (K\dfrac{\partial T}{\partial n})_{III} = q_2$ on surface S_{q2}

(v) $-K\dfrac{\partial T}{\partial n} = h\,(T - T_0)$ on surfaces S_h

and in case of internal duct within the device [20]:

(vi) $-K\dfrac{\partial T}{\partial n} = h^*\,(T - T_0)$ on surfaces S_{h^*}.

where n is the direction normal to S_{qo}, S_{q1}, S_{q2}, S_h, or S_{h^*} as appropriate, T_s is the specified temperature along boundary S_T, T_0 is the ambient temperature, and h and h^* are the convective heat transfer coefficients on the surface S_h and S_{h^*}, respectively.

Using the variational principle for transport phenomena, Hiroaka and Tanaka [21] showed that solving equation (10.1) subject to these boundary conditions is equivalent to finding the function $T(x,y)$ that satisfies the same boundary conditions and minimizes the integral $I(T)$ defined as:

$$I(T) = \int_A \frac{K}{2}[(\frac{\partial T}{\partial x})^2 + (\frac{\partial T}{\partial y})^2]\,dA + \int_A \rho\,C\,(u_x\frac{\partial t}{\partial x} + u_y\frac{\partial t}{\partial y})\,T\,dA$$

$$+ \int_{S_q} q\,T\,dS + \int_{S_h} h\,(\frac{T^2}{2} - T\,T_0)\,dS \tag{10.2}$$

where A and S indicate area and surface integral, respectively, and where the temperature gradients $\partial \bar{T}/\partial x$ and $\partial \bar{T}/\partial y$ are variationally invariant. The minimization of $I(T)$ with respect to the unknown function T is the basis of the finite-element formulation.

Three cases of boundary conditions are considered for the comparison.

Case 1: Dry Cutting. The boundary surface S_h in Figure 10.4 is the surface of the natural convective heat transfer for both of the tool and the tool holder. Since air is a bad heat conductor and the temperatures on S_h are not extremely high, heat losses to the air by conduction, convection, and radiation are negligibly small [22]. Childs et al. [18] showed that within about 1 mm of the cutting edge on the tool flank surface the closeness of this surface to the cooler machined surface can result in conductive heat losses in dry air with mean heat transfer coefficient, h, of about 10^3 w/m^{2}K. Therefore, a small value of h (10 w/m^{2}K) is used on all the external surface S_h except at this small surface, mentioned earlier on the tool flank. The ambient temperature T_0 = 20°C is applied on all external surfaces and T_s = 20°C.

Case 2: Wet Cutting. Childs [18] flooded a water-based coolant directly onto the cutting tool at a rate of 2.5 l min^{-1} in a stream having a diameter of about 10 mm. The value of the convective heat transfer coefficient (h), from the tool to the coolant, was estimated to be about 5×10^3 Wm^{-2} K^{-1}. This value gave temperature contours at the cutting edge well matched for the experimental results. Also, the ambient temperature was T_0 = 20°C and T_s = 20°C.

Case 3: New Cooling Device. In addition to the boundary conditions of case 1, the underside of the cutting insert is attached to a copper seat with high thermal conductivity (385 W/m-K) and a circular duct with 2 mm diameter is constructed in both of the copper seat and the tool holder (Figure 10.1). The thermal contact resistance between the copper seat and the underside of the insert can be neglected by enhancing the contact conductance between their surfaces. This can be achieved by reducing surface roughness and increasing the joint (contact) pressure. Also, an interstitial material such as silicon-based thermal grease, whose thermal conductivity is as much as 50 times that of air, will decrease the contact resistance [23]. The circular duct is subjected to convective heat transfer due to the effect of the cooling fluid followed at high Reynolds number. In the case of using a closed-cycle system with water as a cooling fluid, the estimated value of the convective heat transfer coefficient, h^*, is 10^5 W/m^2-K as calculated in Appendix 10A, T_0 = 20°C and T_s = 20°C. Also, the estimation of the pressure drop and the head loss within the device, based on some given dimensions for the coolant path, is shown in Appendix 10B.

10.4.2 Solution and Results

Using the ANSYS finite-element analysis program an approximate solution of temperature distribution is obtained for each case. For the three cases of loading, the rate of total heat removed by cooling through the tool, Q_{tool}, and the maximum temperatures on the rake and the flank surfaces of the tool are calculated and tabulated as follows. Although the value of Q_{tool} is estimated based on the two-dimensional model, by using the width of cut w as a third dimension, it gives a good indication about the effectiveness of the cooling method used in each case.

Case of Loading	Dry Cutting	Wet Cutting	New Cooling
Maximum Rake Temp. (°C)	712.6	696.5	701.8
Maximum Flank Temp. (°C)	593.8	527.4	560.72
Q_{tool} (W)	13.6	30.5	26.1

The results show a significant amount of the cooling effect of the cutting fluid can be compensated by using the device, the overheated area at the tool tip is reduced compared to the case of dry cutting, and consequently, less wear and longer tool life can be achieved.

Since the device has proven to be analytically successful, the following sections show how quality can be integrated into the design before the actual manufacturing of the device by incorporating customers desires and needs. This is what is called quality function deployment, as discussed next.

10.5   QUALITY FUNCTION DEPLOYMENT

To be able to integrate quality into design, customers' needs and desires must be established and translated into engineering technical requirements. This process is part of quality function deployment (QFD), a tool traditionally used for new product design and development, where customers' needs are deployed into every phase of design and manufacturing. The American Supplier Institute defines QFD as "a system for translating the voice of the customer into appropriate company requirements at each stage from research and product development to engineering and manufacturing to marketing/sales and distribution" [24]. QFD is achieved by cross-functional teams that collect, interpret, document, and prioritize customer requirements to identify bottlenecks and/or breakthrough opportunities. The key to QFD's competitive advantage is its structured application of four strategic concepts [25,26]:

- Preservation of the voice of the customer: ensures that the customer won't be changed in the process
- Cross-functional team: provides input to product realization from all areas of the business.
- Concurrent engineering: allows those parties, such as manufacturing, that have traditionally participated later in the product realization cycle to begin planning earlier, using more accurate information.
- A concise graphical display: presents a picture of the product that links customer needs to product-realization decisions.

In a typical application, the voice of the customer (end user) is captured and translated into product and process design requirements. These requirements are organized in a series of matrices to help the company in moving from product planning through manufacturing planning.

10.5.1   Elements of QFD

QFD has two key elements. These are:

1. *The team:* The QFD team comprises representatives from each of the major work groups involved in a product realization process. Because this is a cross-functional team, each member brings special knowledge about the capabilities and requirements of his or her processes, taking every time in account the customer needs. The team operates by consensus. Thus issues are discussed until a decision is reached that is supportable by each team member.

2. *The house of quality (HOQ):* Akao [27] defined a series of matrices to help guide the QFD process. The first and most frequently applied QFD matrix, also called the "house of quality," translates the customer's need into measurable technical attributes. This graphical display provides a framework that guides the team through the QFD process.

To "build" this house, six basic steps must be followed:

1. *Voice of the customer:* What are the needs and wants of your customers?
2. *Competitive analysis:* In the eyes of your customers, how well is your company doing relative to your competition?
3. *Voice of the engineer:* What are the technical measures that will relate to your customer's needs?
4. *Correlations:* What are the relationships between the voice of the customer and the voice of the engineer?
5. *Technical comparison:* How does your product performance compare to your competition?
6. *Trade-offs:* What are the potential technical trade-offs?

These steps will now be used to analyze the cooling device and evaluate its quality requirements.

10.5.2   Application of QFD

Step 1: Voice of the customer. Since QFD in its simplest definition represents "voice of the customer," customer demands must drive the design of the new product. For example, to build our new cooling device, it is known that the heat generated at the tool–work interface is a major cause of operating inefficiency, excessive tool wear, frequent shutdowns, poor finishes, and other costly problems. To help overcome these problems, the advantages and disadvantages of cutting fluids have been examined thoroughly.

Knowledge of these advantages and disadvantages from the customers can help decide what are the manufacturers', as customers in our example, needs and wants and can help categorizing the cutting fluid effects on the production process into three categories:

- Effects on workpiece material.
- Effects on machine tool and its various component, such as bearing and slide ways.
- Biological and environmental effects.

Because the machine-tool operator is always dealing with the cutting fluid, the effects of contact with the fluid should be of primary concern. In addition to fumes, smoke, and odors, the fluid can also cause several reactions on the skin and various parts of the operator's body. An additional important factor is the consideration of green engineering in our design, that is, considering the effect of the fluid on the environment, particularly in regard to its degradation and ultimate disposal.

To consider these customers' needs, the new cooling device should be compact in size and provide the following:

1. Dry cutting area.
2. Constant and uniform cutting tool.
3. Noiseless operations.
4. No effect on the environment.
5. No effect on the operator.
6. Improved cutting process.

These needs are shown in Figure 10.5, and each corresponds to one of three types of features: basic feature, one-dimensional feature, and excitement feature. Their role in determining customer response to the product differs. Basic features are assumed by the customer as either dissatisfied or satisfied. If they are not met, then the customer is dissatisfied. One-dimensional features will increase customer satisfaction as performance improves. However, there is little excitement generated by these features. Excitement features delight the customer and differentiate a product from its competitors. Over time, excitement features become one-dimensional and one-dimensional features become basic. Each need was evaluated according to the basic (B), one-dimensional (O), and excitement (E) categories, and the results were placed in the Step 1 area of the house of quality shown in Figure 10.5.

Step 2: Competitive analysis. To evaluate the importance and fulfillment of each need discussed in Step 1, customers were surveyed on "how important is this need to you?" and "how well do the other products (cutting fluids in our example) meet this need?" The customers responded using a scale of 1 (low) to 5 (high). The data were compiled and the results of the customers' evaluations were entered into the house of quality of Figure 10.5. In this figure, numerical results are listed in column A while graphical depictions are placed to the right of column G. The importance rating data are placed in column E.

After reviewing the customers' evaluation and importance data, all customers agreed that cutting fluids have harmful effects on the environment, operator, and manufacturing processes. Therefore, the new cooling device should "eliminate" any or all of these effects. Based on the customers' evaluations (column A) and the importance data (column E), a target level for each customer need must be decided

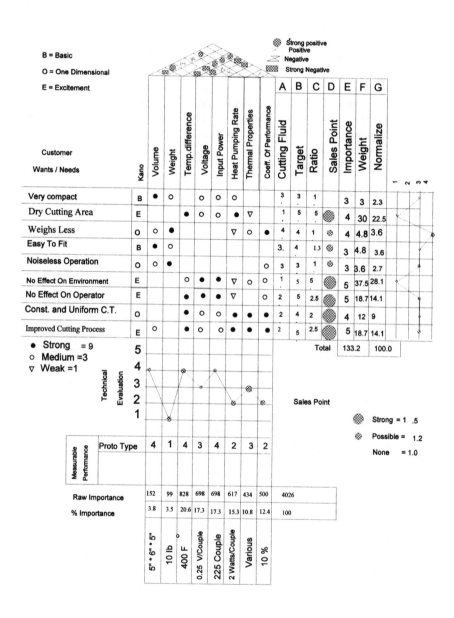

Figure 10.5 House of quality.

based on a scale of 1 (low) to 5 (high) as shown in column B. The target is the level of improvement that is desired after the new product is introduced. Column C presents measures of the relative improvement necessary to meet these targets, computed by dividing the target value (column B) by the current performance (column A).

Special emphasis can be placed on a few customer needs by selecting one of the three values for a sales point (column D). A weight of 1.5 is a very strong sales point; 1.2 is a possible sales point; 1.0 is not a sales point. In reaching a decision as to the degree of sales point, it was taken into account whether the item had been identified in Step 1 as a basic, one-dimensional, or excitement characteristic. It was decided that the needs with the strongest sales points are "Dry cutting tool," "No effect on environment," "No effect on the operator," and "Improved cutting process."

Next, a weighing score (column F) was calculated by multiplying the data in column C (ratio), column D (sales point), and column E (importance). The weights were then normalized by converting to percentages (column G). This allowed a quick view of the relative ranking of each item under customer needs as shown in Figure 10.5.

Step 3: Voice of the engineer. Voice of the engineer is a major key in building the HOQ. Engineers are faced with the challenge of translating the subjective customer statements into objective engineering performance measures. To accomplish this, a QFD team must create a list of technical measurements that are most likely to predict the degree to which a need is fulfilled. The items on this list are engineering parameters that are internal to a company. At least one technical requirement should be identified for each customer need. These technical measures are also referred to as "substitute quality characteristics" because an engineering test is being substituted for what the customer desires. As with the customer needs, the number of technical measures should be kept to not more than 20 to 30. For our cooling device, the following list of technical requirements is generated and placed in the matrix of Figure 10.5:

1. Volume (cm^3)
2. Weight (kg)
3. Maximum temperature difference between the cooling device and the cutting tool ($^{\circ}$C)
4. The applied voltage (V) that maximizes the heat pumping rate (volts)
5. The current (I), which maximizes heat pumping rate (amperes) and the power input (watts)
6. The maximum heat pumping rate ($\Delta T/sec$)
7. Thermal properties
8. Coefficient of performance

Step 4: Correlations. The HOQ developed in Figure 10.5 can now be used to predict customer satisfaction. This can be achieved by asking questions on how well the manufacturers are satisfying their customers' needs and wants. The answer to these questions are discussed by the QFD team until consensus is reached and symbols signifying strong, medium, and weak relationships are used in the house. If no relationship exists, then the matrix space is left blank. These symbols are used to facilitate identifying patterns of the relationships.

As can be seen from Figure 10.5, thermal properties of materials used in building the device (e.g., copper seat) play a very effective role in satisfying the customer needs. Having a good conductive material helps in removing more heat, which, in turn, affects the manufacturing process and the cutting tool. Also, it affects the volume and weight of the device, and consequently it affects the ease of fit and the noise level.

Step 5: Technical comparison. Technical comparison is often accomplished by obtaining specifications from competitors' literature or by measuring actual performance. This ensures that all products are examined under the same environmental conditions with the same testing equipment and may be the only way of collecting data concerning the technical measures identified in Step 4. Since the cooling device is very new in the market, there is no other competitive device to do a technical comparison between them. But to complete this step, a patent search has been conducted to evaluate similar design concepts.

A quick review of Figure 10.5 shows that the prototype is heavy, the heat pumping rate is low, and the input power is high. This indicates that the weight can be decreased by selecting another material with good thermal properties and low density. Additionally, the input power needs to be decreased and the heat pumping rate needs to be increased. This is called "raw importance" and it is calculated for each technical measurement. This raw importance measure combines the importance of the customer's needs and wants (from Step 2) with the strength of their correlations with the technical measures (from Step 4). Numerical weights of 9, 3, 1, and 0 correspond, respectively, to strong, medium, weak, and nonexistent correlations. For each customer demand (d) and technical measurement (t), let S denote the strength of the relationship and let I denote the importance weight (from column E). Then,

$$\text{raw importance of } t = \Sigma_d S_{dt} I_d$$

For example, the raw importance of weight for the cooling device = 3*3 + 9*4.8 + 3*4.8 + 9*3.6 = 99.

The raw importance scores can be converted into percentages. Target values are then selected for every technical measure. Typically a few critical technical measures are targeted for improvement, based on the importance scores, the competitions, and the company's comparison to that competition.

Step 6: Technical trade-offs. At this stage, the roof of the house must be completed. This requires a verified or documented design using new engineering tools such as concurrent engineering. Once this design is established, technical trade-offs can be determined. This simply means examining design features for possible improvements. If improving one design feature can lead to improvement of another feature, then a strong positive correlation exists (depicted by the double circle). If there is a strong possibility that the other feature gets worse, then a strong negative correlation exists (depicted by double x). Empty spaces in the roof of the house indicate no correlations, a single x indicates negative correlation, and a single circle indicates positive correlation. For example, if there is an increase in "volume" accompanied by an increase in "temperature difference" and "heat pumping rate," then there is a positive correlation. If increasing the input power leads to a decrease in the performance, then there is a negative correlation. We must now weigh the trade-offs or, better yet, approach the conflict as an opportunity for a technical breakthrough.

10.6   SUMMARY

Design for quality (DFQ) is becoming more widely used today throughout the world. The implementation of DFQ will continue to help reducing the product introduction cycle time by increasing productivity and quality of the products. One of the tools of DFQ is the quality function deployment (QFD). QFD is a very useful tool that helps the design team to ask the right questions—questions that would be otherwise be overlooked, such as, what are the customers requirements? what are the design requirements? and the most important and most frequently overlooked, what are their relationships? An important application of QFD in the machine tool industry was presented in this chapter. QFD was used to design a new device that would eliminate the biological and environmental effects of cutting fluids. The device would enhance dry cutting processes at higher speeds and feed rates, improve accuracy, and enhance quality and tolerance of manufactured products. In addition, the device was designed with the customer's voice in mind. This helped in producing a new cooling device that is compact in size and is environmentally friendly. This practical application of quality to machining proves that quality plays a major role not only in all efforts of product development but also in continuous improvement.

10.7   BIBLIOGRAPHY

1.   Trent, E.M., *Metal Cutting*, 3rd ed., Butterworth-Heinemann, 1991.
2.   Shaw, M.C., *Metal Cutting Principles*, Clarendon Press, Oxford, 1984.
3.   Niebel, B.W., Draper, A.B., and Wysk, R.A., *Modern Manufacturing Process Engineering*, McGraw-Hill, New York, 1989.
4.   Lindberg, R.A., *Processes and Materials of Manufacture*, 4th ed., Allyn & Bacon, 1990.
5.   Byers, Jerry P., *Metalworking Fluids*, Marcel Dekker, New York, 1994.
6.   Smart, E.F., and Trent, E.M. Coolants and Cutting Tool Temperatures, *Proceedings 15th International Conference M.T.D.R.*, pp. 187–195, 1972.
7.   Kalpakjian, S., *Manufacturing Engineering and Technology*, Addison Wesley, Reading, MA, 1995.
8.   Lofton, H. "Cold Lubricant Misting Device And Method," U.S. Patent 4,919,232, 1990.
9.   Kurimoto, T., and Barrow, G. The Influence of Aqueous Fluids on the Wear Characteristics and Life of Carbide Cutting Tools, *Annals of the CIRP*, 31(1), 1982.

10. Ber, A., and Goldblatt, M. The Influence of Temperature Gradient on Cutting Tool's Life, *Annals of the CIRP*, 38(1), 1989.

11. Jeffries, N. P., "A New Cooling Method for Metal-Cutting Tools," Ph.D. Dissertation, Department of Mechanical Engineering, University of Cincinnati, 1969.

12. Dudly, G. M., "Machine Tool Having Internally Routed Cryogenic Fluid for Cooling Interface Between Cutting Edge of Tool and Workpiece," U.S. Patent 3,971,114, 1976.

13. Meyers, P. G., "Tool Cooling Apparatus," U.S. Patent 3,137,184, 1964.

14. EXAIR Corporation, "Cold Gun Aircoolant System," technical catalogue.

15. Mazurkiewicz, M., Kubala, Z., and Chow, J. Metal Machining with High-Pressure Water-Jet Cooling Assistance—A New Possibility, *Journal of Engineering for Industry*, 111(7), February 1989.

16. Mazurkiewicz, M., "High Pressure Lubricooling Machining of Metals," U.S. Patent 5,148,728, September 1992.

17. *Tool and Manufacturing Engineers Handbook*, 3rd ed., SME, 1994.

18. Childs, T.H.C., Maekawa, K., and Maulik, P., Effects of Coolant on Temperature Distribution in Metal Machining, *Material Science and Technology*, 4, November 1988.

19. Maekawa, K., Ohshima, I., and Suzuki, K. Improvements in Cutting Efficiency of Ti-6Al-6V-2Sn Titaniam Alloy (2nd Report)—Investigations for Reducing Tool Tip Temperature, *J. JSPE*, 59, p. 6, 1993.

20. Maekawa, K., Ohshima, I., and Murata, R., Finite Element Analysis of Temperature and Stresses Within an Internally Cooled Cutting Tool, *Bulletin of the Japan Society of Proc. Engineering*, 23(3), September 1989.

21. Hiroaka, M., and Tanaka, K. "A Variational Principle Transport Phenomena," Memo, Faculty of Engineering, Kyoto University, 30, pp. 235–263, 1968.

22. Tay, A.O., Stevenson, M. G., and de Vahl Davis, G. Using the Finite Element Method to Determine Temperature Distributions in Orthogonal Machining, *Proceedings Institution of Mechanical Engineers*, 188, pp. 627–638, 1974.

23. Incropera, Frank P., *Fundamentals of Heat and Mass Transfer*, 3rd ed., John Wiley and Sons, New York, 1990.

24. American Supplier Institute, *Quality Function Deployment Information Manual*, 1989.

25. Mallon, J.C., and Mulligan, D.E. Quality Function Deployment—A System for Meeting Customers' Needs, *Journal of Construction Engineering and Management*, 119(3), September 1993.

26. Brown, P.G. QFD: Echoing the Voice of the Customer, *AT&T Technical Journal*, pp. 18–31, March/April 1991.

27. Akao, Y. Quality Function Deployment, *Quality Progress*, October 1983.

28. Chapman, A. J., *Heat Transfer*, 4th ed., Macmillan, New York, 1984.

29. Techo R., Tickner R. R., and James R. E. An Accurate Equation for the

Computation of the Friction Factor for Smooth Pipes from the Reynolds Number, *Journal of Applied Mechanics*, 32, p. 443, 1965.

30. Gnielinski V., New Equations for Heat and Mass Transfer in Turbulent Pipe and Channel Flow, *International Chemical Engineering*, 16, pp. 359–368, 1976.

31. Munson, B.R., Young, D.F., and Okiishi, T.H. *Fundamentals of Fluid Mechanics*, John Wiley & Sons, New York, 1990.

32. Billatos, S.B., and Ayad, A. An Innovative Approach to Environmentally Safe Machining, proceedings of the Design for Manufacturability Conference, *ASME International Mechanical Engineering Congress & Exposition*, November 6–11, Chicago, pp. 1–8, 1994.

33. Billatos, S.B., and Abdel-Malak, N. A Design for the Environment Application in Machining. *Proceedings of the SME Fifth World Conference on Robotics Research*, Cambridge, MA, May 1994.

10.8 PROBLEMS / CASE STUDIES

10.1 The engineering department of ABC Tooling Inc. is developing a new tool using the new cooling-device methodology suggested in this chapter. The department is concerned about the type of material necessary to build the tool. Which considerations of the material are required to satisfy the customer needs and assure a long tool life? In your decision, you must understand that there are three possible modes by which a cutting tool can fail in machining. These are (i) fracture failure (the cutting force becomes excessive at the tool tip causing it to fail suddenly by brittle fracture), (ii) temperature failure (the cutting temperature at the tool–workpiece interface becomes excessive, causing plastic deformation and loss of sharp edge), and (iii) gradual wear (cutting edge of the tool starts to wear out gradually, e.g., flank wear, causing inefficient surface finish and less precision and accuracy of the workpiece).

10.2 Contaminated cutting fluids could be a major source of pollution in the metal cutting industry. Three problems can be identified as follows. First, aerosols and/or mists resulting from the use of air as suction or blast from spraying can leave particles suspended in the air. These particles are a hazard to workers through inhalation and generate pollution. Second, ground seepage that occurs with improper holding tanks for the cutting fluids or by simple runoff to the ground could be a safety and fire hazard. Finally, metal particles or shavings throughout the cutting process can contaminate fluids. In order to clean the fluids, these metals have to be filtered out, disposed of, and removed from the workplace for future recycling. Suggest possible solutions to each of these three problems.

10.3 Chip disposal is a problem that is often encountered in turning and other continuous operations. Continuous chips are often generated, especially when turning ductile materials at high speeds. These chips may develop a secondary shear zone, which becomes deeper as tool-friction increases. The type of chip varies depending on the type of material and tool used. But in general, chips are hazardous to the machine operator and damage the workpiece finish. To avoid the problem, some chip breakers have been developed. Company X, which is famous for efficient environmental solutions combined with customer satisfaction, has decided to create the "Envo-Breaker," a special environmentally friendly chip breaker. Using this new piece, the machining is safe for the operation and there is no damage to the final surfaces of the pieces.

a.　Which customer needs does Company X have to take into account?

b.　Describe the voice of the engineer in the design of the "Envo-Breaker."

10.4　In the first weeks of usage of the "Envo-Breaker," designed in problem 10.3, the customers of the Company X were satisfied. The cost of injuries to machines operators due to the chips decreased, and the final surfaces improved in a considerable way. Nevertheless, after some time, the customers realized that large amounts of chips were produced and needed to be removed from the work area. The problem was translated to the customer needs department of Company X, which decided to create a combination of the chip breaker and a collection system. Two types of devices were created: the "Breaker-Vacuum," a combination of the functions of break and vacuum, and the "Breaker-Gravity," a system to break the chip and collect it by allowing gravity to drop the chip on a steel conveyor.

a.　What could be the new voice of customer for this case?

b.　Describe the voice of the engineer for the two types of the device.

c.　Try to build a HOQ for one type of the device.

10.5　Due to the new environmental laws, the chip generated in the machining processes have been declared "solid waste." The government requested that all the metalworking companies reduce the volume of chips by compaction (crushing), prior to hauling away the chips from the manufacturing plant. A survey conducted by the Company X revealed that the crushing program was one of the important needs of the customer.

　　　The president of the Company X, producing the Breaker of problem 10.4, and the senior marketing manager went to the engineering department with the results of the survey. After few weeks of hard work, the engineering department presented new designs of the recycling system, but it didn't satisfy the board of trustees, which was expecting a combination of the "Breaker-Vacuum" or the "Breaker-Gravity" with the recycling system.

a.　What are the completed customer needs?

b.　Which are the new considerations for the device?

c. Try to design a recycling system that combines the three functions:
break–collection–recycle.

APPENDIX 10A—ESTIMATION OF THE CONVECTIVE HEAT TRANSFER COEFFICIENT: FROM COPPER SEAT TO INTERNAL COOLANT

Known:

- Hydraulic diameter of the internal circular duct, $D_h = 2.0$ mm.
- The cooling fluid is water at 20°C with the following properties [28]:

Prandtl number	$P_r = 6.99$
Thermal conductivity	$K = 0.5996$ W/m-K
Density	$\rho = 998.3$ kg/m^3
Dynamic viscosity	$\mu = 1.003 \times 10^{-3}$ kg/m-s
Coolant discharge rate	$Q_c = 5.0$ liter/min

Assumptions:

- Turbulent flow in smooth pipes.
- Fully developed flow.
- The boundary conditions at the wall of the duct are uniform heat flux or uniform wall temperature.
- Single-phase convection in which the maximum coolant temperature inside the duct is lower than the boiling temperature.

Analysis:

Since the discharge rate

$$Q_c = \rho V\ \pi D^2/4 = 5.0/1000 \text{ m}^3/\text{min} \times 998.3 \text{ kg/m}^3 = 4.9915 \text{ kg/min} = 0.08319 \text{ kg/sec}$$

then the coolant velocity inside the duct is V = 26.526 m/sec, and the Reynolds number is $Re_D = \rho\ V D / \mu = 0.528 \times 10^5$.

For fully developed turbulent flow in a smooth circular duct, the Techo et al. correlation [29]:

$$\frac{1}{\sqrt{f}} = 1.7372\ \ln \frac{Re}{1.964\ \ln Re\ -\ 3.8215}$$

has been used to get the flow friction factor $f = 0.005164$. Also, for smooth-walled ducts, Gnielinski correlation [30],

$$Nu = \frac{(f/2) \ (Re \ - \ 1000) \ Pr}{1 \ + \ 12.7 \ (f/2)^{1/2} \ (Pr^{2/3} \ - \ 1)}$$

has been used to get Nusselt number $Nu = 344.47$, and since $Nu = (h^* D) / K$, then $h^* = 10^5$ W/m²-K.

APPENDIX 10B—ESTIMATION OF THE ENERGY LOSSES IN THE COOLING DEVICE

Known:

• The length (l) and the diameter (D) for each portion of the coolant path, as shown in the figure.
• The same cooling fluid properties as mentioned in Appendix 10A.

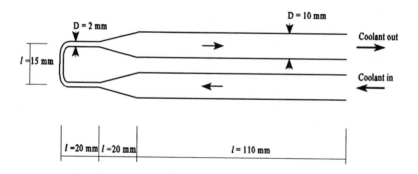

Schematic of the coolant path.

Assumptions:

• Steady incompressible turbulent flow in horizontal smooth circular ducts.
• Fully developed flow.

Analysis:

In the straight portions that have constant diameter D, the pressure drop

$$\Delta p = 4f \frac{l}{D} \frac{\rho V^2}{2}$$

and the head loss h_L between any two sections

$$h_L = \Delta p / \gamma$$

where f is the flow friction factor as calculated before in Appendix 10A, and γ is the specific weight of the water.

Substituting the appropriate values of l, D, V, and f of each portion and summing all the head losses, the total head loss due to the friction will be 20.5 m.

Also, there are minor losses occur due to the gradual change in pipe diameter and the existing of bends. To determine these head losses, the loss coefficients, K_L, are used. Thus,

$$h_L = K_L (V^2 / 2g)$$

where the values of K_L are dependent on the geometry of the components considered and on the fluid properties. These values are estimated from several experimental works gathered by Munson et al. as follows [31]:

- For the conical contraction portion, based on the downstream velocity, $K_L = 0.02$.
- For the 90° bend, $K_L = 0.2$.
- For the conical diffuser, based on the upstream velocity, $K_L = 1.0$.

Substituting with the appropriate values of K_L and V and summing all the head losses, the total head loss due to the minor losses will be 50.9 m. Consequently, the overall estimated head loss due to the major and minor losses in the device will be 71.4 m and the total pressure drop

$$\Delta p = \gamma h_L = 700 \text{ kpa}$$

Thus, the total power loss in the device can be obtained as

$$\mathcal{P}_L = \gamma Q c \, h_L = 58.3 \text{ W} = 0.08 \text{ hp}$$

Also, it should be noted here that this estimation of the power loss is the lower limit of the actual value since the assumption of the fully developed flow in smooth circular ducts cannot be valid in the real situation.

ELEVEN

LIFE-CYCLE ANALYSIS AND MAINTAINABILITY

This chapter introduces the concept of life-cycle design: a design concept that takes into consideration the life of a product during the design stage. Service mode analysis is briefly presented and covers the costs incurred to service a product during its lifetime. It highlights the causes of recurring problems, common in industry, during a product's life cycle. The chapter highly stresses design for maintainability and how this concept is completed prior to final design formulation. The easier a product is to maintain, the more customers are attracted to it and ultimately will purchase that product.

11.1   INTRODUCTION

In the last 25 years growing international competition and increasing consumer demand for higher quality, reliability, and supportability have increased the complexity of product design. This trend has forced U.S. companies to implement quality methods not just in the manufacturing process but also in the design, testing, and support operations of the product. In the past, U.S. companies tended to place a greater emphasis on solving problems after a product had been developed. This created low-quality products, high product development costs, and innumerable delays in getting a product out to market. A single defect that would cost a company less than a hundred dollars to correct if detected during development can end up costing thousands of dollars to diagnose and correct the problem if it is detected in the factory. It can cost even more in terms of loss of reputation or even loss of lives if detection doesn't occur until it reaches the field. On the other hand, it is cost-prohibitive for a company to design for 100% product quality, reliability, and maintainability. Instead, significant cost savings can be realized just by placing emphasis on quality, reliability, and maintainability early in the design phase of the product. This emphasis must be pursued continuously throughout the development, testing, and manufacturing stages of the product. The focus of this section is to provide an overview of product life-cycle analysis, serviceability, and maintainability, a description of the methodologies involved, such as the criteria function, and methods to assess its performance. Throughout this section, it is important to realize that designing for quality, predictability, reliability, and maintainability is a process that involves a multidisciplined engineering approach.

11.2   LIFE-CYCLE DESIGN

The success of any product is directly related to the quality of its design. Getting a high-quality product with minimal cost has become the most important aspect in product design today, and design for maintainability (DFMAIN) has become an important tool in this process. Life-cycle analysis is the study of the total cost of acquiring, operating, and maintaining a product over its entire life. This includes the cost to design, produce, operate, maintain, and even dispose of the product. The objectives of designing to life-cycle cost are to identify the major cost drivers throughout the design process, to perform trade-off analyses of competing design and serviceability approaches, and to estimate a product's total cost over its entire life.

Life-cycle design encompasses the entire product life, from conception to disposal/recycling. The seven criteria of life-cycle design are:

- Company policy
- Life-cycle costs
- Environmental protection
- Resource optimization
- Working conditions
- Product properties
- Ease of manufacturing

These factors constitute the criteria function that sets the environment in which designs are carried out. The criteria function is discussed in the following section, and it consists of all the external influences the designer is subject to when defining the following product characteristics:

- Need
- Design and development
- Production
- Distribution
- Usage
- Disposal/recycling

These factors describe the phases the product "experiences." This concept is an expansion of the current product life-cycle model most manufacturers follow. The differences lie in the fact that all criteria functions are considered important at every stage in a product's life. This is especially important at the need and design/development stages because it is at these stages that most of the product's impact is set with respect to ease of manufacture, ease of assembly, environmental impact, etc. Therefore, this is very much like an extension of concurrent engineering, in that many functions are carried out simultaneously. Instead of design, analysis, purchasing, etc., it is the concurrent inclusion of many issues, or criteria, that is happening.

11.3   THE CRITERIA FUNCTION

The seven proposed issues or criteria to be considered as a framework for product development are only an outline. The particulars of each need are set by the manufacturer to ensure alignment with their objective goals.

Most companies have already included these criteria in their *company policy* to a large degree, but probably not with respect to all phases of a product's life. It is assumed that very few companies have an established policy concerning the safe disposal, or recycling, of their products. The corporate leaders need to set the direction to take, keeping in mind the developing regulatory climate and consumer preferences. This policy then needs to be effectively communicated to the entire organization in order to be successfully implemented.

Life-cycle costs have traditionally not included environmental costs associated with each individual product. In general, environmental costs have been lumped together with general overhead costs. This has masked the contribution of particularly "bad" products, making it difficult to significantly decrease manufacturing wastes. The proposed change would associate all costs, including environmental costs from manufacturing, using, and ultimately disposing (waste handling, treatment, disposal, EPA Superfund litigation, etc.) of each individual product. In this way, a company has the information available to make more intelligent decisions about that product. The regulatory climate in the United States is heading in the direction of mandating this accountability. The companies that take the lead will be able to plan their future accordingly, instead of reacting later.

Resource optimization is just what it sounds like, using the available resources as efficiently as possible. The difference lies in expanding the concept to cover all phases of the product's life, from definition of the need to recycle. The most important parts of this are material and energy flow analysis. In this way, a complete account of the resources used is accomplished. Again, the regulatory climate is headed in this direction, as are consumer preferences. The example that comes to mind is the EPA corporate average automobile fleet mileage requirements.

The *working conditions* concept is expanded to include both product use and recycle/disposal. This has traditionally been thought of in terms of ease of product use for the consumer. A shift in thinking is required to bring this criteria to the forefront of the design cycle. The tools used for the accurate environmental and health assessment of a product are in need of further extensive development work.

Product properties are the traditional focus of the designer's efforts. The optimum will change in response to the added criteria. A CIM environment is the perfect tool to use for monitoring and controlling these properties. There needs to be a tie-in with other accounting tools and databases to allow for a true life-cycle analysis in "real time."

Ease of manufacturing is a result of company policies established within the company's strategic plan guidelines. There will be some changes as the environmental costs for each product become clearer with the development of accounting methods. The environmental properties of a product are included along with its manufacturing and assembly properties.

These seven proposed criteria contribute to the overall enhancement of the environmental protection effort that is so necessary to building a sustainable economy. The net result for the manufacturer will be a more efficient operation due to reduction in plant losses. Traditionally, these losses were not easily isolated and corrected, so they continued to propagate. In today's global marketplace, the level of competition ensures that only the efficient manufacturer will prosper.

To fully adopt each of the seven criteria functions previously mentioned will require further development of those concepts. The overall life-cycle concept is still in the development stage, as are many of the tools. The material and energy flow analysis will require much work simply because much of the pertinent information is not collected in a common database. Some of the development tools are most equitably the responsibility of governmental regulatory agencies.

A corollary facet of DFE is the fact that these designs often are more reliable due to their reduction of the number of parts, but, serviceability may diminish due to the increases in part sizes and complexities. If serviceability and reliability are not examined thoroughly at the design stage, the savings in manufacturing costs using DFE may be eclipsed by the cost increases in warranties and service work. With this in mind, the following section examines a way to build serviceability into a design.

11.4   DESIGN FOR SERVICEABILITY

The goal of serviceability design is simple. As in all designs, customer satisfaction must be maximized while minimizing costs. It is impossible to do both, so a healthy balance must be found. Serviceability is measured as the ease to perform all service-related operations. This is highly likened to the example of automotive design. A more easily and cheaply serviced automobile makes a customer happy. While it is not possible to make a car completely reliable for its lifetime (generally 100,000 miles), some services must be accomplished such as routine maintenance (oil changes, tune-ups, etc.). If the car were service free, it would probably be unaffordable. Therefore, a healthy balance must be met between the following factors: cost of the car, reliability of the car during its lifetime, cost of necessary devices, amount of necessary services, and ease of services. The following list are the factors influencing serviceability for any design [1]:

1. *Reliability* of components and subsystems.
2. *Labor cost* to do necessary services and repairs for the customer (services and

failures), and for the manufacturer (warranty work).

3. *Cost to the manufacturer to stock replacement* parts, tools, and equipment for the life cycle of the product and these costs to the consumer.

4. *Accessibility of components to be serviced.* For example, oil changes would be much more expensive if the engine needed to be removed to do the service.

5. *Availability of necessary parts*, and anything else needed for service. This would apply to the customer as well as the manufacturer. Lack of availability of any of these items increases costs and down time of the product, which affect customer satisfaction.

6. *Mechanic training.* Difficulty in performing services decreases customer satisfaction. This is due to higher costs at shops and customer inability to perform the service oneself. It also increases down time if the customer must take the product to specialty shops.

7. *Customer preferences.* This factor has many variables within each design. It is hard to pinpoint exactly what will satisfy the most potential consumers. A good example would be a fan belt with a 50,000-mile lifetime. The majority of customers may prefer to service/change it twice during the car's lifetime since this service can be scheduled. Even though performing the two servicings requires more money, a customer may prefer it over an unscheduled catastrophic breakdown. These are the preferences and trade-offs that must be considered, examined, and designed for.

8. *Where the product is serviced.* Much like number 6, a customer may prefer to have the choice to do a service at home, thereby saving time and money.

9. *Length and coverage of warranty.* A trade-off exists here also. A longer warranty means higher costs to the manufacturer ,which must be passed along to the consumer. But a shorter warranty means a cheaper and less reliable auto. Customer expectation and competition must be polled here to see what is expected. Then the trade-off must be weighed. The ultimate choice must lead to the greatest customer satisfaction weighing all concerns therein. The warranty cost is determined by the mean time between failures, the mean time to repair, the parts per repair, and the expected product use for the product.

10. *Special tools, repair equipment, and replacement part production.* Standardization of repair parts and tools will ease serviceability. A hard-to-get tool turns a simple repair into a major problem that will create excess cost and lost time.

These factors make it difficult to weigh and judge how to design for drivability because so many customer preferences and cost trade-offs occur. Does doubling the price of an item warrant doubling the mean time between failures? Does reducing the number of parts in a system (DFE goal for cheaper manufacturability) make it more or less serviceable? This is the challenge of the serviceability engineer. Serviceability plays an important role in customer satisfaction because a product that is difficult or expensive to service will make the customer unhappy with the product.

Serviceability can be further subdivided into two main functions: diagnosability and maintainability. Diagnosability is the capability of easily identifying the source of the problem and providing the course of action to correct it. Maintainability is the capability of performing maintenance on routine or regular basis. This function is discussed in detail in the following section.

11.5   DESIGN FOR MAINTAINABILITY

Maintainability is generally defined as the probability that a failed item can be repaired in a specified amount of time using a specified set of resources.   The military publication of "Definition of Terms for Reliability and Maintainability," MIL-STD-721C defines maintainability as the measure of the ability of an item to be retained in or restored to a specified condition when maintenance is performed by personnel having specified skill levels, using prescribed procedures and resources, at each prescribed level of maintenance and repair [2].   These definitions indicate that maintainability is greatly influenced by many variables such as availability of resources and environmental conditions where maintenance is performed.   The following cases illustrate the wide variety of maintenance activities that are performed where maintainability is taken into account [2–7]:

- In electronic instruments that have a large number of interconnected electronic components, maintenance is typically fault-finding by procedure and replacing the damaged components.
- In a large processing plant with a highly automated, 24-hour continuous production, any local failure can stop production or even endanger worker safety.   Therefore the maintenance activity revolves around production availability and worker safety.
- In a car, there is a wide variation of components to satisfy the large variety of customers.   In this case, the car manufacturer builds in maintenance or reduces maintenance work during the design and product development stages to avoid increasing costs.
- In a space station, there is very little availability for maintenance.   In this case, reliability is satisfied by using high-quality components and redundancy.

In many cases, maintainability is more than just another critical element of design. In systems like aircraft and automobiles, a lack of maintainability may cause loss of life. Needless to say, no manufacturer wants its product or design to be associated with death and catastrophe, nor does the manufacturer want to produce a product that fails quite often. In fact, what good is a product that is extremely functional but has a tendency to breakdown or fail every so often?! Therefore, rather than predicting how the equipment will fail, the effort of maintainability is to make manufacturing systems and products that require minimal maintenance and that are easy and inexpensive to fix when they fail. Consequently, manufacturers strive to design maintainability into the products and their manufacturing processes.

Design for maintainability (DFMAIN) is significantly enhanced if, as concurrent engineering (CE) dictates, it is done before the design is formalized. The reason is that design changes during production are very costly. But if DFMAIN is implemented early in the design stage, a great number of benefits would be realized including [8–10]:

1. Longer life of systems and products.

2. Lower operating costs, by performing at high efficiency.
3. Lower unscheduled downtime, by preventing failures.
4. Lower scheduled downtime, by decreasing the time required to perform a particular maintenance task.

An additional and very significant advantage of DFMAIN is the production of *environmentally* friendly products. A product that is easier to maintain can also inherently be a product that is easier to disassemble and dispose off. This meets one of the many new directions to design for the environment [11]. Therefore, to fully realize some or all of the these benefits, some practical maintainability guidelines are next presented. These guidelines are discussed and are followed by a case study to illustrate their applicability. A model that helps in prediction of maintainability is also derived and used for rating DFMAIN characteristics.

11.6 DFMAIN GUIDELINES

Careful design for maintainability can achieve substantial economies in manufacturing. A general rule for DFMAIN is to reduce the possibility of damage to the product and/or equipment during maintenance and servicing or better yet to eliminate the need for maintenance. To assist in establishing a solid foundation for implementing DFMAIN in the early stages of design, a set of standard and organized guidelines is provided. Use of these guidelines can help improve productivity and enhance quality. Solutions to problems are not offered since solutions are unique and continue to require individual judgment and creativity. Instead, the guidelines are based on sequences of key maintainability issues or concepts that may help to ensure a thorough understanding and control of the intended work. Understanding and control will help in generating decisions that may assure reliability, schedules, leverage, and future strength. Figure 11.1 shows how these guidelines can be integrated together during the life cycle of the product and they are described next [2–14].

- *Keep the functional and physical characteristics as simple as possible.* Complexity of design has a direct bearing on maintenance and production cost. The design must consider the following three skill levels for maintenance personnel: (I) the skill of the owner/operator of equipment such as automobiles, lawn mower, etc., (ii) the skill of the technician or professional service engineer who can perform specialized tasks, and (iii) the skill of the major maintenance crew members who can perform maintenance at the depot level for major machine components.
- *Design for recyclability by selecting appropriate materials that suit the intended production method as well as the functional design requirements.* Avoid materials that are difficult to clean, are susceptible to solvents commonly used for cleaning, or have poor adhesion characteristics when bonding is required. Label materials for ease of identification at the recycling stage.
- *Design for the most economical production methods* taking into consider-

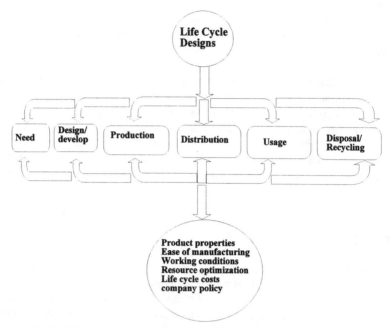

Figure 11.1 Life cycle concept of product design.

ation factors such as the weight ratio of the raw material to the finished product, surface finish requirements, etc.

- *Design for ease of assembly and disassembly with a minimum number of parts.* A product that is easily assembled and disassembled is also easily maintainable. An example of this is calculators that used integrated circuits to create a 30:1 reduction in parts. An additional benefit is that this has reduced the cost of a calculator from $1000 to $150.

- *Design snap-together parts and minimize assembly directions and part variations.* This would minimize the need for special tools and preclude damage of equipment and product during maintenance and servicing.

- *Design subassemblies to be able to test and maintain at that level and not at the final assembly.* Fixing a problem or making different models is accomplished more easily if modular assemblies are used. Modular designs also reduce manufacturing costs by creating group technologies.

- *Build from the base up to eliminate problems from assembly tolerance stackup.* Make major assemblies self-standing, including the base, and take the center-of-gravity location into account during design phase. Self-standing units ease maintenance on-line and eliminate the need for special maintenance areas.

- *Provide unobstructed access.* Make sure that there is enough room for the assembly of parts and any required tooling during assembly, and that the product is easily maintainable and its parts can be easily repaired or replaced.

- *Provide easily read and understood control and labeling systems.* Simple words and symbols that are familiar to all personnel help in reducing improper operations and the consequences, which can range from a few minutes of downtime to the loss of life.
- *Minimize cables, wiring and any tangled parts.* If needed, provide adequate viewing and hand access and use codes or labels for easy identification.
- *Minimize fasteners and adhesives* to avoid the need of additional tooling, the possibility of fasteners being defective and jamming, and adhesive dispensing problems.
- *Locate test, power, pneumatic, or other interfaces on one side only for easy access from one direction and simple tooling and assembly setup.* Locate handling areas and gripping features at balancing points.
- *Determine how a product should be manufactured.* Use standard processes, tools, tests, etc., wherever practical. Use simple tools to lower the expensive capital investments and to ease the tooling setups.
- *Designs must include calibration, standards, etc., for accurately maintaining process performance.* Evaluate suspected problems on primary and secondary operations for a given product rapidly and efficiently. This could be achieved by including measurement equipment and monitors in the design for areas such as fluid flow or where temperature is a concern.
- *Eliminate the need for lubrication systems through the use of self-lubricating parts.* If this is not possible, provide automated lubrication system where the only maintenance required is to clean the system and add lubricant.
- *Design for robust quality by considering the environment in which the product will be in service.* Hazards such as vibration, electrical noise, humidity, and temperature must be accounted for in this guideline.
- *Design for maintainability with safety in mind.* This is a very significant and important issue, especially in a hazardous environment or in instances where maintenance must be performed while the equipment is still in service. The design should minimize weight and awkwardness in handling of parts that must be removed. It should provide for safety guards to prevent contact with moving parts and high temperature or electrical equipment, and fool-proof safety locks that prevent movement of equipment that may not be in the focus of the person performing maintenance.

11.7 RATING DESIGN FOR MAINTAINABILITY

A rating worksheet was developed as shown in Figures 11.2 and 11.3. The two figures rate simple maintenance procedure of two comparable car models from two manufacturers: A and B. This simple task is selected for its familiarity to readers at any level of expertise. To perform the task of oil change, the owner/operator would drain the used oil, remove the oil filter, install a new filter, add fresh oil, and check the oil level. Performing this task on model B requires lying on your back and extending

	# of operations	wt	A 1	B 2	C 2	D 2	E 2	F 1	G 1	H 2
Access	ONE HAND ACCESS	2			1	1				1
	TWO HAND ACCESS	1		1		1	1	1	1	
	TOOL ACCESS (Y/N)	1			1					
	CATWALK	3								
	LADDER PLATFORM	4								
	CRAWL SPACE	3								
	KNEELING	2								1
	LYING	2		1	1					
	STANDING	1				1	1	1	1	
	SUSPENDED (HARNESS)	5								
	PART VISIBLE (Y/N)	1			1	1				1
Door/ Plate	HINGED	1								1
	REMOVED	2								
	SCREWED/BOLTED	2								
	SNAP OPEN	1								
Weight	OVER 45 LBS	3								
	BETWEEN 10 - 45 LBS	2								
	UNDER 10 LBS	1		1	1	1	1	1	1	1
	HANDLES OR GRASPS (Y/N)	1		1	1					
	WINCH/CRANE	3								
Tools	HANDS	1		1			1	1	1	1
	HAND TOOLS	2			1					
	POWER TOOLS	3								
	SPECIAL TOOLS	4				1				
Fastener	SEALANT	3			1					
	EPOXY	3								
	BOLTS/SCREWS/THREADED	2								
	SNAP-FIT/QUICK-CONNECT	1								
	GUIDES (KEY WAY, PINS) (Y/N)	1			1	1				
Monitoring	GRAPHIC	1							1	
	DIAL/LEVEL INDICATORS	1							1	
	"IDIOT" LIGHT	2	1							
	COLOR CODED (Y/N)	1								
	SCHEDULED MONITORING	2	1							
	AUTOMATED MONITORING	1								
	CLEANING	1								
Fault ID	AUTOMATED	1								
	MEASUREMENT COUPLING	2								
	VISUAL	1								
	DEDUCTIVE	2								
	SPECIAL TEST EQUIPMENT	3								
Symbols/ Instruct	RECOGNIZE SYMBOLS (Y/N)	1		1	1	1	1			1
	RECOGNIZE WORDS (Y/N)	1		1	1	1	1			1
	INSTRUCTIONS/ DIAGRAMS (Y/N)	1								
Environment	COMFORTABLE ENVIRONMENT (Y/N)	1								
	ABOVE/BELOW FLOOR LEVEL	2								
	CONFINED SPACE	3								
	RESPIRATOR NEEDED	4								
	ELECTRICAL CONTACTS	3								
	SAFETY GUARDS (Y/N)	1			1	1	1		1	
	EQUIPMENT IN SERVICE	3								
Skill	OWNER/OPERATOR	1	1							
	SKILLED TECHNICIAN	2								
	DEPOT LEVEL	3								
Lube	SELF-LUBRICATING	1								
	TIME/USAGE CHANGE	2	1							

Key

A	General
B	Place/Remove Oil Catch Pan
C	Remove/ Replace Plug
D	Remove/Install Oil Filter
E	Remove/ Replace Cap
F	Add Oil
G	Check Oil Level
H	Open/Close Hood

Total Weighted Score: 130

Figure 11.2   Model B.

	# of operations	wt	A 1	B 2	C 2	D 2	E 2	F 1	G 1	H 2
Access	ONE HAND ACCESS	2			1					1
	TWO HAND ACCESS	1		1		1	1	1	1	
	TOOL ACCESS (Y/N)	2								
	CATWALK	3								
	LADDER PLATFORM	4								
	CRAWL SPACE	3								
	KNEELING	2								
	LYING	2		1	1					
	STANDING	1				1	1	1	1	1
	SUSPENDED (HARNESS)	5								
	PART VISIBLE (Y/N)	2								
Door/	HINGED	1								1
Plate	REMOVED	2								
	SCREWED/BOLTED	2								
	SNAP OPEN	1								
Weight	OVER 45 LBS	3								
	BETWEEN 10 - 45 LBS	2								
	UNDER 10 LBS	1		1	1	1	1	1	1	1
	HANDLES OR GRASPS (Y/N)	2		1	1					
	WINCH/CRANE	3								
Tools	HANDS	1		1			1	1	1	1
	HAND TOOLS	2			1	1				
	POWER TOOLS	3								
	SPECIAL TOOLS	4								
Fastener	SEALANT	3			1					
	EPOXY	3								
	BOLTS/SCREWS/THREADED	2			1	1	1			
	SNAP-FIT/QUICK-CONNECT	1								
	GUIDES (KEY WAY, PINS) (Y/N)	2			1	1				
Monitoring	GRAPHIC	1								
	DIAL/LEVEL INDICATORS	1							1	
	"IDIOT" LIGHT	2	1							
	COLOR CODED (Y/N)	2								
	SCHEDULED MONITORING	2	1							
	AUTOMATED MONITORING	1								
	CLEANING	1								
Fault ID	AUTOMATED	1								
	MEASUREMENT COUPLING	2								
	VISUAL	1								
	DEDUCTIVE	2								
	SPECIAL TEST EQUIPMENT	3								
Symbols/	RECOGNIZE SYMBOLS (Y/N)	2		1	1	1				
Instruct	RECOGNIZE WORDS (Y/N)	2		1	1	1				
	INSTRUCTIONS/ DIAGRAMS (Y/N)	2								
Environmen	COMFORTABLE ENVIRONMENT (Y/N)	2								
	ABOVE/BELOW FLOOR LEVEL	2								
	CONFINED SPACE	3								
	RESPIRATOR NEEDED	4								
	ELECTRICAL CONTACTS	3								
	SAFETY GUARDS (Y/N)	2			1	1	1		1	
	EQUIPMENT IN SERVICE	3								
Skill	OWNER/OPERATOR	1	1							
	SKILLED TECHNICIAN	2								
	DEPOT LEVEL	3								
Lube	SELF-LUBRICATING	1								
	TIME/USAGE CHANGE	2	1							

Key

A	General
B	Place/Remove Oil Catch Pan
C	Remove/ Replace Plug
D	Remove/Install Oil Filter
E	Remove/ Replace Cap
F	Add Oil
G	Check Oil Level
H	Open/Close Hood

Total Weighted Score: 115

Figure 11.3 Model A.

an arm holding a socket wrench to a very difficult to reach and poorly visible area of the oil pan to remove the plug, drain the oil, and then replace it. To remove and install the oil filter, a filter wrench is normally used; however, clearance to fit the wrench around the head of the filter was not provided so the filter had to be removed and installed using two hands. This created considerable difficulty in reaching the torque requirements and was very uncomfortable for the person performing the maintenance. Finally, the oil inlet and oil-level stick were not marked as such and were difficult to locate. In contrast to model B, reaching the oil pan plug to drain the oil from model A is also performed lying on your back; however, the plug is relatively easy to reach and visible. The oil filter is well placed so that both a filter wrench and the hands of the person performing the task can easily reach around it. Finally, both the oil inlet and oil-level stick are clearly marked.

These tasks were rated and results showed that model A received a rating of 115 and model B a rating of 130. Predictably, model B scored higher and would therefore be considered more difficult to change the oil for as compared to model A. The ramifications for these results are that the maintenance task performed on model B, as opposed to model A, will incur higher costs, and decreased customer satisfaction, which will affect sales. The worksheet also makes clear the shortcomings of the design by allowing the designer to identify in which areas a lower score could be reached through modification. Ease of manufacturing is a result of company policies established within its strategic plan guidelines. These will be some changes as the environmental costs for each product become clearer with the development of accounting methods. The environmental properties of a product are included along with its manufacturing and assembly properties.

11.8   SUMMARY

In this chapter, a number of practical maintainability guidelines have been presented. These guidelines can be used to improve production and assembly lines, monitor process selection, and establish criteria for quality-control. Although they were applied to simple applications, these guidelines can be tailored to provide specific solutions to engineering problems. They also raise many key questions that would lead to a better understanding, control, and improvement of the production environment.

11.9   BIBLIOGRAPHY

1.   Gershenson, J., and Ishii, K.   Life-Cycle Serviceability Design   in *Concurrent Engineering: Tools & Techniques* , ed. A. Kusiak, Wiley & Sons, New York, 1993.

2.   Klement, M.A. Design for Maintainability, *Concurrent Engineering: Tools & Techniques,* ed. A. Kusiak, Wiley & Sons, New York, 1993.

3.   Holmberg, K., and Folkeson, A. *Operational Reliability and Systematic Maintenance*, Elsevier Applied Science, 1991.

4.   Miller, W.G. Planning & Controlling New Product Start Up: A Practical Approach, *Proceedings, SYNERGY*, pp. 144–146, 1984.

5.   Moss, M. *Designing for Minimal Maintenance Expense: The Practical Application of Reliability and Maintainability*, Marcel Dekker, New York, 1985.

6.   Priest, J. *Engineering Design For Productibility and Reliability*, Marcel Dekker, New York, 1988.

7.   Smith, D. *Reliability and Maintainability in Perspective*, Halsted Press, 1985.

8.   Ayres, R.U. Complexity, Reliability, and Design: Manufacturing Implications, *Journal of Manufacturing Review*, 1(2), 26–35, 1988.

9.   Billatos, S.B. Guidelines for Productivity and Manufacturability Strategy, *Journal of Manufacturing Review*, 1(3), pp. 164–167, October 1988.

10.   Billatos, S.B., and Grigely L.J. Functional Requirements Mapping as a Framework for Concurrent Engineering, *International Journal of Concurrent Engineering: Research, and Applications (CERA)*, Vol. 3, 1993.

11.   Billatos, S.B., and Nevrekar, V.V. Challenges and Practical Solutions to Designing for the Environment, *ASME National Design Engineering Conference*, Chicago, March 14–17, 1994.

12.   Boothroyd, G., Dewhurst, P., and Knight, W. *Product Design for Manufacture and Assembly*, Marcel Dekker, New York, 1994.

13.   Stool, H.W. A Design Backwards Approach to Product Optimization, *SME Conference on Simultaneous Engineering*, 1987.

14.   Suh, N.P., Bell, A.C., and Gossard, D.C. On an Axiomatic Approach to Manufacturing and Manufacturing Systems, *ASME Journal of Engineering Industry*, 100(2), 1978.

11.10   PROBLEMS / CASE STUDIES

11.1   Barbara Stanwick has been asked to evaluate the performance of the design and development process for the aircraft engine controller. The evaluation was called for after the company had failed to deliver on some changes that its best client had asked for, to include additional functional improvements, in order to compete with the other airlines. These additional changes led to cost overruns and designs that were difficult to test or to integrate with the existing design. Many different engineers were brought in to work on the project at different stages but none stayed throughout the length of the project. Six months later none of the functional requirements had been met and the client was threatening to take its business elsewhere. In analyzing the problem, Barbara found that the problem resulted from placing more emphasis on the hardware design and not enough on the software design for the controller until failures in the software design also led to delays in the whole project. She found that the program was not well documented because the engineers that had worked on the project had already moved on before the documentation started. Also the program functions were not modularized or well structured, making it difficult to integrate the new requirements without a great deal of rework. Finally, the software was not fully tested before moving from one phase to another. In fact, no one was checking if the software was meeting any of its requirements until the last stage, after which many problems were found.

a. Why is proper documentation important for the software design? When should documentation start?

b. Why is it important to modularize the software?

c. The software phases include software requirements, requirements analysis, preliminary and detailed design, coding and unit testing, final audits, and product delivery. At which phases should the software be reviewed?

11.2 A robotics manufacturer for automated assembly lines is looking into improving the serviceability and maintainability of the robot. It conducted different tests and found that the components could be grouped into three categories based on product life: 1–3 years, 4–7 years, and 8–12 years. The technology life of the different components differed from these, with the shortest one lasting 1 year and the longest lasting 5 years for the most significant component of the robot. The company wants to group the components into modules in order to allow it to produce a family of products based on the inclusion or exclusion of these modules.

a. What is the difference between product and technology life?

b. What components should be grouped into modules in order to improve serviceability and maintainability?

c. Which components should be highly accessible based on reliability studies and importance to operations?

d. How does grouping the components into modules improve the ability of the company to upgrade and service their family of products?

11.3 A manufacturer of refrigerators wants to be an industry leader in green manufacturing. It is looking into offering costumers the service of taking back old refrigerators and selling a newer model at a discounted price. How can the manufacturer make the effort pay for itself using green manufacturing? Describe some of the changes that would need to go into the product in order to make this happen.

11.4 A motor manufacturer has hired you to analyze the cost of warranty failures in order to decide whether to continue to warranty a product. In order to define the general repair cost you should evaluate labor, facility, replacement, test equipment, shipping, travel, administrative, and training cost. The mean time between failures is as expected, but as you look at the cost of the failures that occur you are surprised to find that it is 75% higher than expected based on the cost data. You dig deeper and find that 20% of the cost can be attributed to the misuse of the product. The rest of the cost is the result of lack of supportability. It was found that the team designing the motor was also the team responsible for responding to customer questions and problems. As a result the team was falling behind both in coming up with new products with the improvements to deal with the customer concerns and in responding to customer concerns.

a.   What suggestions would you make to the manager to improve supportability of the product?
b.   What can the company do to reduce the misuse of the product?
c.   What sort of warranty should the company offer for the new motor models?

11.5   Most household appliances and exercise machines are assembled by the customer. Select two similar products from two different manufacturers and run a scoring system like the one presented in this chapter. Record your findings and describe the problems associated with each product. Also suggest some design solutions using the design for the environment and design for maintainability guidelines presented earlier. Examples include bicycles, computer centers, entertainment centers, chairs, dinning tables, etc.

TWELVE

MULTIATTRIBUTE UTILITY ANALYSIS OF MUNICIPAL RECYCLING

This chapter discusses the concept of municipal recycling. It helps the interested reader in analyzing the various aspects including the methodology and applications of municipal recycling. Also included within this chapter is a short checklist that helps in determining the optimal recycling program within a certain region of population. This checklist takes into account a wide gamut of inputs ranging from the physical makeup of the waste to area legislation, and from public education to budgetary support. The chapter also assists in understanding the subject of municipal recycling and aids in implementing a waste management program best suited for a particular area.

12.1   INTRODUCTION

During the 1960s and the 1970s, the evaluation of complex issues (including uncertainty) was addressed primarily through the use of concepts such as decision trees and weighted expected value analytical routines. To deal with the quantitative and qualitative aspects of decisions in the 1990s, management scientists have developed processes and analytical routines based on "utility" theory concepts. Utility functions help the decision maker to model attitudes toward risk taking and uncertainty and to make value trade-offs among conflicting objectives in a way that is both practical and conceptually sound. "Algebraic formulation methods (including nonlinearity of utility functions)" allow compact representation of decision models that are too complex to handle with decision trees [1].

A. W. Drake argued for using values as the criterion for selecting between alternatives when the stakes at risk are relatively low and the impact is of short duration [2]. We would all be willing to pay $1 for a lottery ticket where we have equal chances of winning $20 or losing $5. However, what would the "utility" be if the ticket cost $25,000 and we had equal probabilities of winning $100,000 or losing $50,000? A decision maker's attitude toward risk is addressed in multiattribute decision analysis and utility theory (MAUT), which utilizes the concept of certainty or certainty equivalent. This is defined as the amount of value that is equally preferred to an uncertain alternative. This is one of the key concepts utilized in MAUT models. If certainty equivalents are known (or posited) by all affected parties to a complex problem, it is easy to ascertain the most preferred alternative. It is the one with the highest (or lowest depending on the type of problem) certainty equivalent. The development of utility theory incorporating this concept was expanded in the late 1970s and 1980s to include more complex value functions that addressed strength of preferences between pairs of consequences. Decision scientists are now more skilled not only at assessing utility functions but also at structuring objectives and in measuring achievement in terms of these objectives [3].

Finally, the utilization of computer programs has greatly assisted analyses in this area and has improved the value of structuring in both the quantitative and qualitative aspects of decision analysis. The use of the computer has allowed us to become more skilled at structuring conflicting goals and objectives, assessing utility functions, creating alternatives, and performing sensitivity analysis on the qualitative as well as the quantitative evaluation criteria. The computer has helped to enable better and improved communication between participants in the decision processes (the stakeholders).

12.2   RELATED RESEARCH AND MOTIVATION

During the past 6 or 7 years there has been a significant increase and improvement in research utilizing MAUT. Corner and Kirkwood published a comprehensive review of more than a hundred applications covering a broad range of topics including energy, manufacturing, services, and health care [4]. Extensive work has been conducted in such areas as the criteria applicable to the siting of a major facility and the shipment of spent nuclear fuel [5,6]. MAUT theory is starting to be applied to life-cycle engineering and is being used to address the environmental impact of a product from "birth to earth" from a manufacturing perspective [7,8]. These efforts have addressed environmental issues in a tertiary fashion. They have not directly looked at the social and political circumstances of the economics of recycling and reclamation activities.

Recycling and related services have become a significant issue within states such as Connecticut where adequate landfill is being rapidly depleted and offshore dumping is no longer permitted. Environmental considerations are being legally enforced. Changes in personal habits and patterns of disposability are being mandated predicated on far reaching concerns and conjecture. It has become politically "in" to be pro environment. The public, forced to pay for far-reaching environmental concerns, has not openly embraced this change. At the confluence of this volatile political and legal quagmire is the local municipality that must administer an effective, efficient recycling program. Effective and efficient in this context refer to being cost-effective and in compliance.

The decision that environmental impact outweighs economic impact for a particular material in a given locality has not been the subject of open and honest debate in many of our cities and towns. MAUT tools have gained wide acceptance through their success in breaking down traditional barriers to communication between groups with conflicting concerns. The solid waste disposal issue poses a serious problem within our society as we deal with legislation designating responsibility for environmental impact.

12.3   METHODOLOGY

We must "internalize the extremities" of solid waste disposal. This requires the structuring of the objectives of the various stakeholders in the municipal recycling decision problem. R. L. Keeney addresses the need to identify very precisely what is critically important to each of the constituent parties to the decision process [6]. This requires interaction among the impacted parties and specificity around what each hopes to accomplish both quantitatively and qualitatively in reaching decisions supporting an optimal recycling program. Keeney describes in detail elicitation techniques for identifying these key goals and objectives. He introduces several key concepts. One concept is discerning "ends" goals from "means" to an end. Less public exposure to toxic material could be a goal from an environmental group, but the underlying objective is to minimize the health impact to the public. A second concept that is crucial to the application of MAUT theory in structuring values is the application of measures or measurements to each of the objectives. These criteria are essential in order to provide a basis for making the value trade-offs for the mathematical development of the utility function. Finally, objectives and measurement criteria must be as specific as possible in framing the overall decision context. For example, a refuse disposal goal might be to minimize the impact to the natural environment. The question "what environmental impacts need to be minimized" would result in a clarification of the objective and provide a better basis on which to measure environmental damage. Development of the goals and measurements hierarchy constitutes the initial phase in delineating problems of this nature. It is absolutely imperative that all stakeholders understand and agree on the goal hierarchy during the initial step.

The basic processes for developing, quantifying, and evaluating the objectives are as follows:

1.   First, define the measurements and attributes for each objective in terms of certainty and/or uncertainty, specifying what is goodness or badness on a criterion, defining how this measure is derived (is it a point estimate, a nonlinear function, a step function) or determining if the measure is more qualitative/subjective (if so how do we view/assess the subjectivity). Scales need to be developed with descriptions and algorithms.

2. Next, the utility function needs to be developed based on these attributes and measurements. There are several ways to accomplish the process. The analytic hierarchy process (AHP) may be employed. This is the simplest of the analytic routines currently utilized by decision scientists. Basically, this involves making pairwise choices between individual criteria or categories of criteria. Based on stakeholder input using AHP, linear algebraic techniques are employed to compute the relative weighting of the various criteria. An aggregate utility function is then derived. This is a relatively simple, straightforward technique and is not utilized in the development of our municipal recycling algorithm. There are significantly more powerful routines available. The software we are using to analyze the municipal recycling decision is called Logical Decisions and has embedded in it higher order routines that maximize the aggregate expected utility of all parties to the recycling decision and deal with the inherent uncertainties more analytically.

The process utilized in the recycling application asks the stakeholders to describe their preference for one attribute compared with another. The software coaches the individual through a rating or ranking comprehension index between the two attributes until the two measures become "equal." All attributes are paired in this way. Interactions between attributes are similarly equated by the software. Based on each stakeholder input on the trade-offs, weighting functions are developed analytically across the model. The attribute weighting functions are combined into an overall utility function for the problem in question.

3. The final step in analytically structuring our recycling problem is the determination and evaluation of potential materials to recycle. Some of these materials may be mandated recyclables. However, the decision maker can still analytically determine the best mix of services for accomplishing the transport of recyclables. Issues such as the cost of curbside pickup compared to resident drop-off could potentially be analyzed even where recycling is mandated utilizing this model.

There are other materials that could be recycled that are not mandated. Economically it may make sense to recycle some of these materials because of cost avoidance (the "saved" cost of landfill or incineration). The objective of this work is to analytically determine the optimal mix of materials for a municipal recycling program. Each material must be "graded" on our attribute/measurement hierarchy, and these quantitative and qualitative assessments must then loaded into Logical Decisions to optimize the municipal recycling program.

12.4 APPLICATION

A combination field study and technical data and information survey was utilized to develop the municipal recycling model. Material for several of the measurements was researched from technical journals. The finalized hierarchy of goals and measurements is presented in Figure 12.1.

The overarching goal at the municipal level is to administer the best possible recycling program for the residents. To accomplish this task the manager must optimize the mix of recyclable materials and services to the community, address political issues, and improve compliance (legalities).

The optimal mix of materials and services is a function of the economics of the decision(s), the ease of use, educating the public, budget flexibility, the potential for unbundling the service or the recyclable material (such as charging for pickup and disposition of old refrigerators), etc.

The economics issue at the local town or municipality level is primarily a "cost avoidance" decision (for every ton of material recycled the town saves "tip" fees of $55/ton, the cost of landfill and/or incineration). The measures applicable to the goals supporting municipal recycling are discussed later. One of these measures, the "cost avoidance index" calculation, was developed for use as a decision-making algorithm for recyclers. Basically, this algorithm is a marginal cost calculation: the avoided cost of either/both landfill and incineration plus any revenues received less pickup/collection costs. This index could be used by municipal recyclers as a means of quantifying the recyclability of a given material.

The maximization of services is primarily associated with improvements in pickup methodology and/or improved participation. Improved participation is measured by the "ease of use." The determinants of curbside recycling (and its economics and efficiencies) were addressed by J. Glenn in his paper on "Efficiencies and Economics of Curb-Side Recycling" [9]. Voluntary versus mandatory participation considerations were analyzed by Folz and Hazlett [10]. These measures as well as additional determinants such as recycling educational consideration are utilized to optimize the "improve participation" goal [11].

To measure quantitatively the overall performance on the "maximize service" goal a "service goal" index is proposed. This index is a function of the cost savings, the cost avoided or foregone, the expected utility improvement (increase in the "diversion rate"), and the anticipated population participation in the new program or service.

Finally, in order to have the best recycling program the manager must maximize the politics of the program and meet minimal legal compliance requirements [12]. These goals and their concurrent measures are combined in the recycling model to derive an overall utility function for the municipal recycling decision. Based on the utility function, three hypothetical materials were reviewed as potential candidates for a municipal recycling program. These were polystyrene (PS), scrap metal, and waste paint. These three materials were chosen for their differing technical properties, the existence or lack of legislation, their absence or inclusion in local programs, and the differing problems these materials pose for pickup and disposal. In short, these are good controversial choices that differ widely in their potential for recycling. The results of the analysis of these three materials are presented in Figures 12.1–12.6.

12.5   RESULTS

The overall hierarchy of goals and measurements for the "best recycling program" is presented in Figure 12.1. Based upon this matrix and the attributes of the three selected potential recyclable materials shown in Figure 12.2, polystyrene would be the optimal material to recycle. Figure 12.2 is a summary of the overall utility ranking of

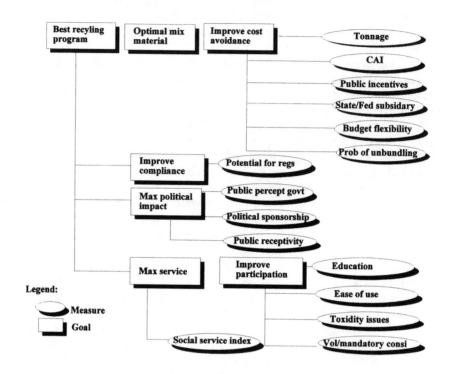

Figure 12.1 Goals and measures of the proposed scheme.

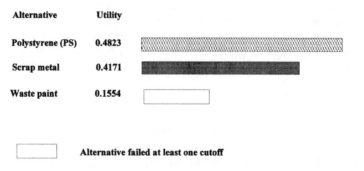

Figure 12.2 Ranking for best recycling program goal for preference set NE.

the three materials based on the weighting factors developed via the process and analytical routines employed. Polystyrene is the easiest to dispose of, is politically a better choice (people perceive "plastics" as bad for the environment), is available in sufficient quantities, and the economics are the best of the three materials.

Waste paint is the least desirable material to recycle. It requires drop-off by residents, special handling, contractor licensing, and the list goes on. It is therefore more costly, and the fact that it is toxic does not outweigh the economic considerations as far as the public is concerned. These conclusions and a sensitivity analysis graph are presented in Figures 12.3–12.6.

Figure 12.3 shows the utility graph for the measure "cost avoidance." As one would expect, as cost avoidance improves for a given material, the utility on this measure increases at an increasing rate. Figure 12.4 presents the factors and weights in the model based on stakeholder input. Table 12.1 and Figures 12.5 and 12.6 present the following:

- A sensitivity graph of the alternatives based on cost avoidance (the vertical line indicates the overall goal weighting percentage).
- A chart of the overall utility of polystyrene versus waste paint (and "why" waste paint was the lowest of the three materials).
- A bar graph of how scrap metal rated on the four major goals (the width of the bar indicates the relative weighting proportion of the goal in the overall decision).

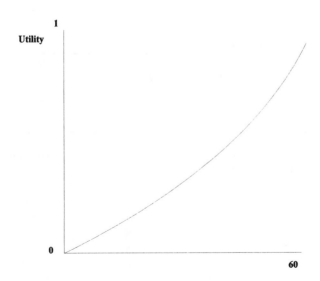

Figure 12.3 Cost avoidance index (× $1000).

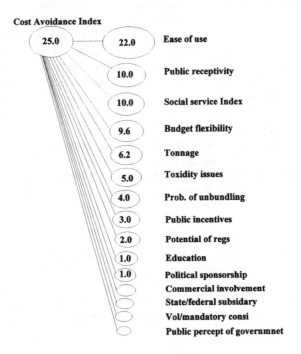

Figure 12.4 Factors and weights used in the model.

Table 12.1. Overall Utility and Total Contribution

Overall Utility for	Polystyrene (PS) (Alt1)			= 0.4823
	Waste Paint (Alt2)			= 0.1554
	Difference			= 0.3269
Measure	Alt1 Level	Alt2 Level	%Contribution to Difference	Total Contribution
Ease of Use	1	6	67.3	0.22
Public Receptivity	1	3	26.5	0.08654
Budget Flexibility	0	3	-16.1	-0.05253
Toxidity Issues	1	0	15.3	0.05
Social Service Index	27.52	4.84	7.0	0.0229
Potential for Regs	3	1	3.1	0.01
Political Sponsorship	1	10	-2.8	-0.009
Commercial Involv.	3	1	-1.5	-0.005
Tonnage	10	5	0.6	0.001892
Cost Avoidance Index	0.734	0.242	0.3	0.001111
Other			0.3	0.0010

12.6   OPTIMAL RECYCLING PROGRAM CHECKLIST

The overall goal of recycling programs can be categorized into four major categories:

- Optimize mix of recycled materials.
- Maximize political impact.
- Maximize recycling services.
- Improve compliance with state and federal regulatory statutes.

A scoring scheme is developed for each of these categories as described below. The scores for this scheme take into consideration many important characteristics such as flexibility, availability, effectiveness, attitude, and cost. They could be used to evaluate the capability and willingness of any federal, state, or municipal authorities to adapt an environmental program such as municipal recycling. This scheme could be modified to reflect other national or local programs.

Optimize Mix of Recycled Materials (Improve Recycling Cost Avoidance).
- Tonnage (potential tonnage of a given material available).
- Public incentives availability (i.e. bottle, can recycling).
- State/federal subsidies available (yes/no).
- Cost avoidance index calculation (avoided cost of either/both landfill and incineration plus revenues less pickup/collection costs). This is a marginal cost calculation of a given potential recyclable material.

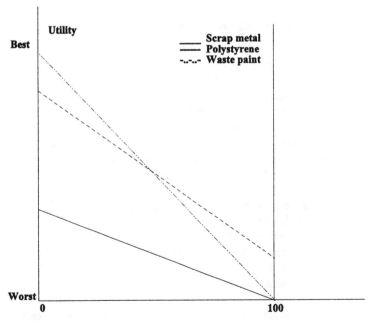

Figure 12.5   Percentage of weight on cost avoidance index measure.

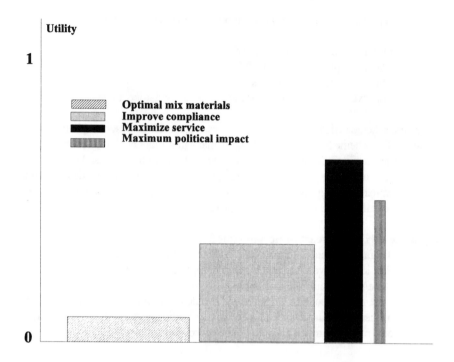

Figure 12.6 Goal member utilities for scrap metal for best recycling program.

- Budget flexibility measure:
    - (0) No budget flexibility.
    - (1) Budget constrained.
    - (2) Moderately constrained.
    - (3) Potential flexibility.
    - (4) Very flexible.
- Probability of unbundling services (i.e. charge for pickup of old appliances, sell special bags for newsprint, etc.) (Prob. %).

Maximize Political Impact of Recycling/Trash Disposal Program.
- Public attitudes/receptivity measure:
    - (1) Proactive, involved.
    - (2) Moderate, single issue support.
    - (3) Disinterest.
    - (4) Moderate distrust.
    - (5) Active opposition.
- Governmental perceptions (attitude toward municipal government):
    - (1) Government is perceived as effective, strong, fair, and representative.

    (0) Municipal authorities will be replaced during next election. This is a scaled measure.

- Political sponsorship for the recycling of a potential material or the improvement of a service. Scaled from 0 to 10, with 0 being no support and 10 being an active commitment.

Maximize Recycling Services. Social service index measure. A measure of the quality of a proposed/current service. The measure is a function of the cost saved, avoided or foregone, the expected utility improvement (increase in the diversion rate), and the anticipated population participation.

Improve Overall Participation.

- Voluntary/mandatory compliance measure (voluntary yes/no).
- Ease of use measure:
    - (1)   Curbside pickup, no special handling or separation required.
    - (2)   Curbside pickup, special bagging required.
    - (3)   Curbside, special handling.
    - (4)   Curbside pickup.
    - (5)   Drop-off at transfer station required.
    - (6)   Scheduled drop-off.
- Education measure:
    - (1)   Public education required through the use of brochures, radio, commercials, endorsements, posters, etc.
    - (0)   No additional educational requirement.
- Commercial involvement measure. Can industrial, commercial organizations and companies be induced to participate either through influence, education, coercion, or outright taxation/penalization actions?
    - (1)   Companies are currently recycling this material.
    - (2)   Recycling could be accomplished through application of the appropriate actions.
    - (3)   Commercial establishments will not comply and the municipality cannot enforce recycling statutes.
- Toxicity considerations. Recycling involves the handling, storage, and transportation of toxic materials. Scaled as a 1 or 0. A 1 means no toxicity issues and a 0 means that a special contractor, special handling, facilities, and licensing are required and there are drop-off/pickup issues.

Improve Compliance. This is the potential for or existence of regulation.

- (1)   Federal, state, or local regulations regarding this material are currently in existence and the local municipality is not in compliance.
- (2)   Regulations are in effect but compliance is marginal or not being adhered to, observed, or compiled properly.
- (3)   More stringent standards are expected to be enacted within the next 12 months and the material under question will be regulated.
- (4)   More stringent standards and a 1 to 3 year planning time horizon.

(5)   No regulatory action is currently anticipated regarding the disposition/regulation of this material.

12.7   SUMMARY

Municipal recyclers face multiple issues when dealing with the disposal and recycling of waste and refuse.   There are the obvious issues concerning the economics of recycling as well as local budget and compliance constraints.   There are also more "qualitative" problems such as the resistance to recycling initiatives, misunderstanding of goals and objectives of recycling, and many other "social" issues.   There are quantitative data, facts, political issues, and concerns to contend with.   All of these issues and concerns must be addressed when establishing an effective municipal recycling program.

Multiattribute decision analysis and utility theory (MAUT) is a method of addressing complex problems and decisions that involve conflicting or contradictory quantitative and qualitative information while at the same time involving all interested parties in the decision process.   The methodological rigor involved in MAUT will be applied to the problem of municipal recycling and the choices/trade-offs that must be addressed in offering the optimal mix of services and recycling options within a municipality.   Issues such as constituent participation, political considerations, toxicity, and economics should be incorporated when developing a comprehensive utility model.

12.8   BIBLIOGRAPHY

1.   Kirkwood, C.W., An Overview of Methods for Applied Decision Analysis, *Interfaces*, 22(6), pp. 28–39, 1992.
2.   Drake, A.W. *Fundamentals of Applied Probability Theory*, McGraw-Hill, New York, 1967.
3.   Keeney, R.L., and Raiffa, H. *Decisions with Multiple Objectives*, Cambridge University Press, 1993.
4.   Corner, J.L., and Kirkwood, C.W. Decision Analysis Applications in the Operations Research Literature, 1970–1989, *Operations Research*, 39, 206–219, 1991.
5.   Keeney, R.L. *Siting Energy Facilities*, Academic Press, New York, 1980.
6.   Keeney, R.L., Structuring Objectives for Problems of Public Interest, *Operations Research*, 36(3), 396–405, 1988.
7.   Burke, D., Beiter, K., and Ishii, K. Lifecycle Design for Recyclability, *ASME Conference on Design Theory and Methodology*, 1992.
8.   Thurston, D.L. Internalizing Environmental Impacts in Design, *Design for Manufacturability, ASME*, DE-Vol.67, 107–113, 1984.
9.   Glenn, J. Efficiencies and Economics of Curbside Recycling, *BioCycle*, 30–34, July 1992.
10.   Folz, D.H., and Hazlett, J.M. Public Participation and Recycling

Performance: Explaining Program Success, *Public Administration Review*, 51(6), 526–532, November/December, 1991.

11. Everett, Jess W., Jacobs, T.L., and Peirce, J.J. Recycling Promotion Strategies: Statistics and Fuzzy-Set Comparisons, *Journal of Urban Planning and Development*, 117(4), 154–167, December 1991.

12. Khator, R. Recycling: A Policy Dilemma for American States, *Policy Studies Journal*, 21(2), 210–226, 1993.

12.9   PROBLEMS / CASE STUDIES

12.1   Source reduction refers to methods of minimizing total disposed volume by eliminating it at the source. Material recyclers now use human labor to sort materials. This is one reason that recycled materials are more costly than virgin materials. Consider ways to automate or mechanically assist the sorting process to improve the yield for a given unit of labor. New sensors might be used to discriminate materials. Example: Infrared (IR) optical spectra of various plastics might be different enough to distinguish materials for guiding pick sorting robots.

12.2   For the situation of problem 1, propose new changes to the product in the design stage that would provide rapid automated material identification for future sorting operations. Example: Make the current plastic recycling ID codes machine readable, maybe with some type of bar code. This must be cheap enough for manufacturers to adopt it without resistance. It might be possible to mold a barcode just below the existing ID code. This code may need to be included on multiple sides of the part to increase the probability that it would be visible on a conveyor belt, and the current triangular ID code might also be changed to a more unique shape to assist machine vision recognition systems. Also consider marking all materials—not just plastics.

12.3   Reinforced plastics provide high structural performance. They have the potential to improve the durability of some products due to corrosion resistance but are difficult or impossible to recycle. Are there ways to chemically reduce and extract the plastic content (depolymerization) for recomposition as near virgin material? What are the economics as compared to raw chemical feed stocks? Could this work for nonreinforced material?

12.4   Survey municipal recyclers for problems and preferences with handling recycled materials. What types of real-world complications exist? What affects costs? What affects the quality of recycled material?

12.5   It will often be impossible to eliminate contaminant materials such as batteries, PC boards, etc., from products. Manufacturers may ultimately be liable for recycling these items (this is already happening in Germany, for example). Consider technical issues such as design features to minimize the cost of compliance. Also consider economic issues. Would a free-market recycling infrastructure be feasible? Third-party reprocessors would accept recycling liability from manufacturers for a fee, and would be required to prove financial solvency similar to the insurance or banking industry. They might also be considered as public utilities. They would then reprocess returned items (in perpetuity) on a more intelligent basis, from recycling plans developed by the manufacturer (and kept on file by the reprocessor). Consumers may need to face the prospect of deposit fees, and reprocessors would probably pay interest on the deposit, especially in cases of lengthy latency periods.

This concept sounds complex and expensive, but consider: Landfill tipping costs are rising, so the cost difference compared to landfill disposal will diminish in time. Third-party reprocessors would relieve manufacturers from future obligations, thus avoiding risk penalties for smaller companies. Foreign companies would release disposal obligation by paying (what amounts to) a disposal tax in the market region. This could create a whole new industry based on "disassembly lines," armed with specific information on exactly what they are reprocessing, thus avoiding the need to test or refuse questionable items. This information-rich infrastructure would require a concurrent engineering environment.

a.   Address the feasibility in terms of economic and environmental issues, liability, and incentives to reduce the level of solid waste through recycling and reuse.

b.   Develop a proposal for a working concept of a free-market recycling infrastructure. Develop design features that could be implemented to minimize cost of compliance.

c.   Determine the pros and cons of a free-market recycling infrastructure. Evaluate current legislation, government/business recycling programs, and public sentiment to determine the potential for a recycling infrastructure (e.g., RECLAIM and ERRA programs).

d.   Use a scoring scheme similar to the one presented in Section 12.6 and evaluate your proposal. Can you promote and sell your proposal to government and/or local businesses?

12.6   Agriculture is a serious source of environmental damage that results from heavy usage of chemicals in the form of pesticides, herbicides, and fertilizers. Large-scale farms are characterized by soil depletion and erosion and contamination of surface and underground water supplies. Newly developed "small cell" oriented farming methods have demonstrated high yields without the use of synthetic chemicals, but are labor-intensive. Can cellular manufacturing methods be applied to agriculture to reduce labor costs? Develop a scoring scheme similar to the one presented in section 12.6 to compare large-scale farms with cellular manufacturing methods.

ACRONYMS

ABS   (Acrylonitrile-Butadiene-Styrene) a thermoplastic–thermopolymer that is primarily made up of three petroleum-derived monomers, called acrylonitrile, butadiene, and styrene.

AD   (Axiomatic Design) a design theory that provides a logical basis for the decomposition of functional and design requirements and supports the concept of a formal language for requirements definition.

AHP   (Analytic Hierarchy Process) a simple analytical routine that is currently utilized by decision scientists. It basically involves making pairwise choices between individual criteria or categories of criteria in multiattribute utility analysis.

BAT   (Best Available Technology) refers to the adaptive nature of current regulations, which require that the most advanced and economically feasible technology available be used for pollution minimization.

CE   (Concurrent Engineering) the tools and working environment that cover the entire life cycle of a product where functional performance and resource requirements of hypothetical design choices can be readily predicted with satisfactory precision. Further, it should provide a flexible structure for evaluation of information to clarify the decision-making process from design details through strategic levels.

CFC   (Chlorinated Fluorocarbon) volatile organic compound used in refrigeration, cleaning solvents, and many other applications. Scientists have linked CFCs to phenomena like global warming and ozone layer depletion.

CIM   (Computer Integrated Manufacturing) is a misnomer if taken literally in today's environment. CIM for the enterprise (lab, plant, company) provides the tools, through the use of hardware and application programs for the generation and integration of engineering and manufacturing data to achieve the most effective design and efficient manufacturing of a product.

DFA   (Design for Assembly) the philosophy and practice of incorporating the functional and design requirements of a product at the early design stage to ensure ease and economy of assembling.

DFD   (Design for Disassembly) the philosophy and practice of using assembly methods and configurations that allow for cost-effective separation and recovery of reusable components and materials.

DFE   (Design for the Environment) the use of design practices where the environmental effects of product function and manufacture are optimized in the design stage. It is a design process that must be considered for conserving and reusing the earth's scarce resources; where energy and material consumption is optimized, minimal waste is generated and output waste streams from any process can be used as the raw materials (inputs) of another.

DFM   (Design for Manufacturing) the philosophy and practice of integrating product design and production resources at the early design stage to ensure ease and economy of manufacturing through the use of automated design and manufacturing tools such as computer-aided design (CAD), computer-aided manufacturing (CAM), etc..

DFMA   (Design for Manufacturing and Assembly) the implementation of all DFA and DFM concepts at the product design stage to ensure ease and economy of manufacturing and assembly.

DFMAIN   (Design for Maintainability) an approach used at the early design stage to measure the ability of a product to be retained in or restored to a specified condition when maintenance is performed using specified following maintainability guidelines.

DFR   (Design for Recyclability) an infrastructure where products can be accepted at the end of their useful life by efficiently breaking them down and then recycling the individual products for use in other processes. In general, DFR is a design process in which a product's environmentally preferable attributes—recyclability, disassembly, maintainability, refurbishability, and reusability—are treated as design objectives where the product's performance, useful life, and functionality are maintained, and even improved.

DFX   (Design for X) a general design strategy where X implies a specific form of optimization, such as zero defects or automated assembly.

It simply refers to any of the design strategies such as design for manufacturing, design for assembly, etc.

DP (Design Parameter) a definition of the design attributes or characteristics conceived to satisfy the functional requirements of the product.

ECM (Environmentally Conscious Manufacturing) the development and application of cleaner, more efficient manufacturing methods to achieve the goals of green engineering.

EOP (End-Of-Pipe) a term that refers to the treatment of harmful by-products after they have been created. This term also refers to the commonly known definition of pollution control.

EV (Electric Vehicle) an electric-powered vehicle that can be considered either renewable or nonrenewable depending on how the electricity powering the vehicle was produced. A vehicle powered with energy from a hydroelectric plant would be using a renewable source of energy. A vehicle powered with energy from coal plant would be using a nonrenewable source of energy.

FR (Functional Requirement) a description of the desired performance level and functions of the product. It is an expression of what the product should achieve.

HDPE (High-Density Polyethylene) a nontoxic thermoplastic material that maintains high resistance to acidic media. It exhibits good toughness at high and low temperatures, and superior electrical and insulating properties.

HOQ (House of Quality) a series of matrices that graphically help guide the quality function deployment process in translating the customer's need into measurable technical attributes.

HSW (Hazardous Solid Waste) a common by-product of manufacturing processes. Examples of these include fuels, cleaners, adhesives, sealants, inks, and certain metals and alloys. They are so classified if they exhibit any hazardous characteristics such as flammability, reactivity, corrosivity, or toxicity:

JIT (Just-In-Time) a manufacturing approach intended to reduce (eliminate if possible) waste in inventory and setup time.

KB (Knowledge Base) contains information on how to improve products by studying the characteristics of each part that makes up the product and the required production processes. It contains rules and data that are used for inference purposes.

KBES (Knowledge-Based Expert System) a specialized computer program that exhibits the same level of problem-solving skills as an expert (person who is very good at solving specific type of problems) for a narrow problem. It embodies knowledge and reasoning capabilities that allow it to draw quality conclusions comparable to those drawn by an expert.

LDPE (Low-Density Polyethylene) a thermoplastic used in the production of grocery bags. It exhibits properties of toughness, high impact strength, low brittleness temperature, flexibility, chemical resistance, low permeability to water, and excellent electrical properties.

MAUT
(Multiattribute Analysis and Utility Theory) a theory that addresses decision makers' attitude toward risk. It utilizes the concept of certainty or certainty equivalent. This is defined as the amount of value that is equally preferred to an uncertain alternative.

NHSW
(Nonhazardous Solid Waste) is generated from numerous manufacturing processes in such forms as metal cutting scrap, rejected stock material, rejected product, and end-of-life product disposal. It includes industrial as well as municipal waste, and is the only category that does not generally define waste as pollution.

PBT
(Polybutylene) a plastic known for its good physical properties, excellent wear, and chemical resistance. PBT can be used to improve the chemical resistance of a polymer.

PE
(Polyethylene) sometimes referred to as polythene, a plastic that contains the ability to withstand low temperatures while retaining its ability to be flexible. It has low water absorption and has an excellent resistance to chemicals. This may be the main reason why polyethylene is so popular in the food and beverage industry.

PET
(Polyethylene Terephthalate) a thermoplastic that displays good tensile and mechanical strength, and is quite resistant to creep, deformation, and exposure to flame. Its superior resistance to impact makes it a safe alternative to glass bottles and containers.

PP
(Polypropylene) a thermoplastic derived from petroleum and made from carbon and hydrogen. It is a highly versatile material. Its impact resistance and flexional strength make it suitable for chair shells, suitcases, television cabinets, toys, etc. Although it has a waxy feel like polyethylene, it cannot be scratched easily. It exhibits good chemical resistance and is easily pigmented.

PPS
(Polyphenylene Sulfide) a thermoplastic that possesses several attractive material properties that makes it ideal for cookware, automotive, and electrical products. Its high melting point and thermal stability are useful in environments with extreme temperatures. PPS is extremely resistant to acidic media and sunlight.

PS
(Polystyrene) a plastic known for its low thermal conductivity, low density, low cost per volume, and energy consumption. Examples of products include thermal insulation for building, ice chests, coffee cups, egg cartons, meat trays, protective packaging, floating devices, and loose fill packaging.

PUR
(Polyurethane) a thermoset that can have a large range of properties depending on the composition of its basic components. Semirigid PUR is characterized by having good tensile strength and high elongation properties. Rigid PUR has a particularly low thermal conductivity. These types are commonly joined with

reinforcements such as fiberglass to increase stiffness and/or heat resistance. It is used in many applications including automotive interior components, refrigerator and freezer insulation, and wind deflectors.

PVC   (Polyvinyl Chloride) a thermoplastic; the simplest structure of most thermoplastics is polyethylene. It maintains a low combustibility, excellent weather resistance, and good dimensional stability. PVC is used as piping, pipe fittings, pumps, sinks, outdoor furniture, bathtubs, electrical wire coating, etc.

QFD   (Quality Function Deployment) a tool traditionally used for new product design and development, where customers' needs are deployed into every phase of design and manufacturing.

RCRA   (Resource Conservation and Recovery Act) a public law that regulates the treatment, transportation, storage and disposal of solid and hazardous waste.

RIM   (Reaction Injection Molding) an injection molding process where a mixture of two or more reactive fluids is forced into the mold cavity. Chemical reactions take place rapidly in the mold and the polymer solidifies, producing a thermoset part. Major applications are automotive bumpers and fenders, thermal insulation of refrigerators and freezers, and stiffness for structural components.

SAN   (Styrene-Acrylonitrile) a strong material with very good chemical resistance produced when styrene is copolymerized with acrylonitrile alone. If SAN is polymerized with butadiene in the form of fine particles, the material's chemical resistance improves and the impact resistance increases.

SMA   (Service Mode Analysis) a procedure used to put a cost on keeping a product functioning for its full life. SMA attaches a price to fix each malfunction that a product may experience using its operational lifetime. It then adds up all of the failures and obtains a total lifetime service cost. It analyzes the malfunctions of the entire system from the customer's point of view.

SPC   (Statistical Process Control) a statistical analysis that is used to control and improve the quality of the products and their manufacturing processes.

TQM   (Total Quality Management) a system used to produce a quality product without relying on feedback concerning defects detected during inspections.

VOC   (Volatile Organic Compound) vapor that is released during painting and cleaning processes. It has been designated by scientists as harming the environment and depleting the earth's ozone layer.

INDEX